Dermatologic Research Techniques

Edited by
Howard I. Maibach, M.D.

CRC Press

Boca Raton   New York   London   Tokyo

Library of Congress Cataloging-in-Publication Data

Dermatologic research techniques / edited by Howard I. Maibach.
     p.   cm. -- (CRC series in dermatology)
  Includes bibliographical references and index.
  ISBN 0-8493-8373-0
  1. Dermatology--Research--Methodology. 2. Skin--Diseases--Animal models. I. Series.
  [DNLM: 1. Dermatology--methods. 2. Research Design. WR 20 D435 1995]
RL79.D47 1995
616.5'027--dc20
DNLM/DLC
for Library of Congress   95-22841
                                                   CIP

CRC Series in
DERMATOLOGY: CLINICAL AND BASIC SCIENCE
Edited by Dr. Howard I. Maibach

The CRC Dermatology Series combines scholarship, basic science, and clinical relevance. These comprehensive references focus on dermal absorption, dermabiology, dermatopharmacology, dermatotoxicology, and occupational and clinical dermatology.

The intellectual theme emphasizes in-depth, easy to comprehend surveys that blend advances in basic science and clinical research with practical aspects of clinical medicine.

Published Titles:

Bioengineering of the Skin: Water and the Stratum Corneum
Peter Elsner, Enzo Berardesca, and Howard I. Maibach

Bioengineering of the Skin: Cutaneous Blood Flow and Erythema
Enzo Berardesca, Peter Elsner, and Howard I. Maibach

Bioengineering of the Skin: Methods and Instrumentation
Enzo Berardesca, Peter Elsner, Klaus P. Wilhelm, and Howard I. Maibach

Hand Eczema
Torkil Menne and Howard I. Maibach

Health Risk Assessment: Dermal and Inhalation Exposure and Absorption of Toxicants
Rhoda G. M. Wang, James B. Knaak, and Howard I. Maibach

**Handbook of Mouse Mutations with Skin and Hair Abnormalities:
Animal Models and Biomedical Tools**
John P. Sundburg

Pigmentation and Pigmentary Disorders
Norman Levine

Protective Gloves for Occupational Use
Gunh Mellström, J.E. Walhberg, and Howard I. Maibach

Forthcoming Titles:

Bioengineering of the Skin: Skin Surface Imaging and Analysis
Klaus P. Wilhelm, Peter Elsner, Enzo Berardesca, and Howard I. Maibach

The Contact Urticaria Syndrome
Arto Lahti and Howard I. Maibach

Dermatologic Research Techniques
Howard I. Maibach

Handbook of Contact Dermatitis
Christopher J. Dannaker, Daniel J. Hogan, and Howard I. Maibach

Human Papillomavirus Infections in Dermatovenereology
Gerd Gross and Geo von Krogh

The Irritant Contact Dermatitis Syndrome
Pieter G.M. van der Valk and Howard I. Maibach

Skin Cancer: Mechanisms and Relevance
Hasan Mukhtar

The Editor

Howard I. Maibach, M.D., is Professor of Dermatology at the School of Medicine, University of California, San Francisco.

Dr. Maibach is a graduate of Tulane University, and has served as Intern at the William Beaumont Army Hospital and as Fellow at the Hospital of the University of Pennsylvania. Dr. Maibach received his training in clinical dermatology at the University of Pennsylvania and, thereafter, embarked on a distinguished career at the University of California, San Francisco.

He has received numerous honors and awards including an honorary doctorate from the University of Paris-Sud and, recently, his peers, research fellows, and students gathered at a conference in Basel specifically convened to celebrate Dr. Maibach's 65th birthday. His professional memberships are numerous and include the International, North American, and European Environmental Contact Dermatitis Research Groups. Dr. Maibach is the author, co-author, or editor of over 1000 research papers and 40 reference texts.

Preface

This book provides an overview and in-depth evaluation of the rapidly evolving world of dermatologic research models. The volume updates the previous four in the "Models in Dermatology" series (Maibach, H. and Lowe, N., *Models in Dermatology*, Karger, Basel and New York). Taken together, they provide a comprehensive approach to the choice of technique for a given area as well as how to interpret the data.

This slim volume and its predecessors offer a hightly time-efficient passkey to this complex cutaneous universe.

We appreciate the kind assistance of the authors in referring, not only to their publications, but also to previously unpublished data.

We would appreciate suggestions from the readers as to how to make these "Models" summaries more useful to the ever-progressing worlds of dermatologic research.

Howard I. Maibach, M.D.

Contributors

Ulla Ågren
Department of Anatomy
University of Kuopio
Kuopio, Finland

Peter H. Anderson
Department of Dermatology
Marselisborg Hospital
University of Aarhus
Aarhus, Denmark

Enzo Berardesca
Department of Dermatology
University of Pavia
Pavia, Italy

Gordon Duncan
Department of Toxicology
ProCyte Corporation
Kirkland, Washington

Irene Dunwiddie
Department of Toxicology
ProCyte Corporation
Kirkland, Washington

Jan Faergemann
Department of Dermatology
Sahlgrenska University Hospital
University of Gothenburg
Gothenburg, Sweden

Gerard J. Gendimenico
Department of Dermatology Research
  and Drug Discovery
Johnson & Johnson Consumer Products
  Worldwide
Skillman, New Jersey

Thomas C. Glenn
Department of Dermatology
University of California, Irvine
Irvine, California

John M. Haigh
School of Pharmacy
Rhodes University
Grahamstown, South Africa

Conrad Hauser
Department of Dermatology
Hospital Cantonal Universitaire
Geneva, Switzerland

Michael Hertl
Department of Dermatology
University of Cologne
Cologne, Germany

Harm HogenEsch
Department of Veterinary Pathobiology
School of Veterinary Medicine
Purdue University
West Lafayette, Indiana

Xiaoying Hui
Department of Veterinary Pathobiology
School of Veterinary Medicine
Purdue University
West Lafayette, Indiana

Bernard Kalis
Department of Dermatology
Hospital Sebastopol
University of Reims
Reims, France

Lloyd E. King, Jr.
Department of Medicine
Vanderbilt University Medical Center
Nashville, Tennessee

Olle Larko
Department of Dermatology
Sahlgrenska University Hospital
University of Gothenburg
Gothenburg, Sweden

Howard I. Maibach
Department of Dermatology
University of California Hospital
San Francisco, California

Jerry L. McCullough
Department of Dermatology
University of California, Irvine
Irvine, California

Hans F. Merk
Department of Dermatology
University Hospital
Aachen, Germany

James A. Mezick
Johnson & Johnson Wound Healing
  Technology Resource Center
Skillman, New Jersey

Kai M. Muller
Department of Biological Structure
University of Washington School of
  Medicine
Seattle, Washington

Doris Niederau
Department of Dermatology
University of Cologne
Aachen, Germany

Shelley Packard
University of Wisconsin
Madison, Wisconsin

Leonard M. Patt
Department of Toxicology
ProCyte Corporation
Kirkland, Washington

Marc Paye
Toxicology Laboratory
Colgate-Palmolive
Milmort, Belgium

Claudine Piérard-Franchimont
Department of Dermatopathology
University of Liege
Liege, Belgium

Gérald E. Piérard
Department of Dermatopathology
University of Liege
Liege, Belgium

Eric W. Smith
Department of Pharmaceutics
School of Pharmaceutical Science
Rhodes University
Grahamstown, South Africa

John P. Sundberg
Department of Pathology
The Jackson Laboratory
Bar Harbor, Maine

Christian Surber
Departments of Dermatology and
  Pharmacy
University Hospital
Basel, Switzerland

Malgorzata Sznitowska
Department of Pharmaceutical
  Technology
Medical Academy of Gdansk
Gdansk, Poland

Markku Tammi
Department of Biomedical Engineering
Cleveland Clinic Foundation
Cleveland, Ohio

Raija Tammi
Department of Anatomy
University of Kuopio
Kuopio, Finland

Erika D. Timpe
Department of Toxicology
ProCyte Corporation
Kirkland, Washington

Ronald E. Trachy
Department of Toxicology
ProCyte Corporation
Kirkland, Washington

Anna-Liisa Tuhkanen-Martikainen
Department of Anatomy
University of Kuopio
Kuopio, Finland

Hideo Uno
Primate Center
University of Wisconsin
Madison, Wisconsin

Pieter G. M. van der Valk
Department of Dermatology
University Hospital
Nijmegen, The Netherlands

Dominique Van Neste
President
Skinterface
Tournai, Belgium

Els B. Stam-Westerveld
Department of Dermatology
University of Groningen Hospital
Groningen, The Netherlands

Ronald C. Wester
Department of Dermatology
University of California Hospital
San Francisco, California

Table of Contents

1 The Guinea Pig as a Model for Topical Dermatopharmacology

Thomas C. Glenn and Jerry L. McCullough

TABLE OF CONTENTS

I.   INTRODUCTION

For decades, the albino guinea pig has been a useful tool in dermatologic research. From the classic Draize skin sensitization model to UV radiation testing, the guinea pig has revealed important pharmacologic and toxicologic information.[1-3] Within the past decade a new strain of euthymic guinea pig was developed that was essentially hairless. This development allowed researchers distinct experimental advantages compared to the albino guinea pig, the foremost being the lack of the need to depilate the animal of its dense fur. Depilation alone can act as a confounding factor in skin experiments by possibly damaging the stratum corneum or protective layer of the skin, thus affecting absorption of topically applied drugs or irritants.

The purpose of this study is twofold: first to investigate an array of physical and chemical paradigms in guinea pig skin. These studies include both classical tests such as irritation to chemical agents and erythema to UV radiation. The second purpose of the study is to compare cutaneous pharmacological responses between albino and hairless guinea pigs.

II. MATERIALS AND METHODS

A. ANIMALS

Male or female Hartley albino and hairless guinea pigs (350 to 500 g) were purchased from Charles River, Inc. All animals were housed and fed *ad libitum*. For albino guinea pigs, hair was first shaved with Oster clippers and then depilated with Nair®. Except where noted, six animals from each group were used in each experiment.

B. TISSUE CHARACTERISTICS

Epidermal thickness: Epidermal thickness was measured with a Zeiss microscope and micrometer. Three biopsies were measured from three dorsal skin areas on each animal.

Epidermal turnover: Six hairless and six albino guinea pigs were injected intraperitoneally with [³H]thymidine (specific activity 20 Ci/mmol). After 1 h guinea pigs were anesthetized and 4-mm punch biopsies were taken from the dorsal skin and were processed for autoradiography. Biopsies taken were on days 1, 3, 5, 7, 10, and 14 after initial isotope injection.

Stratum corneum turnover: A 5% dansyl chloride (Sigma) ointment was made and applied to the dorsal skin of hairless and albino guinea pigs. A Hilltop chamber was secured above the ointment and left in place for 24 h. Guinea pigs were examined on progressive days for fluorescence with UV light. The day when fluorescence disappeared was recorded.

In vitro percutaneous absorption: Two groups of hairless and albino guinea pigs were used; a non-Nair® group and a second group that had been treated with Nair®. Dorsal skin was removed from euthanized animals and trimmed of excess underlying fat. Sections of skin approximately 4 cm² were mounted on Franz diffusion chambers (Crown Glass) that were filled with isotonic saline and 4% bovine serum albumin and equilibrated to 37°C. A 1% solution of hydrocortisone in 60:40 ethanol:water was added topically to the skin. Samples were taken at 4 h intervals up to 48 h and analyzed for [³H]hydrocortisone.

C. IRRITANT STUDIES

The dorsal surface of albino and hairless guinea pigs was divided into eight quadrants. Of the following drugs 0.05 ml was added daily for 4 to 8 days to the dorsal surface with a positive displacement pipet: water, 7% salicylic acid, 10% formaldehyde, 5% lactic acid, 5% benzalkonium chloride, and, in a separate experiment, four concentrations (0.0005, 0.005, 0.05, and 0.1%) of retinoic acid. Sites were scored on a 6-point scale at 1, 4, and 8 days as follows: 0, no erythema; 1, very slight; 2, slight; 3, moderate; 4, severe; 5, very severe; 6, necrosis.

D. DELAYED CONTACT HYPERSENSITIVITY

Anesthetized guinea pigs were given intradermal injections of 0.1% dinitro-2,4-chlorobenzene (DNCB). Seven days later DNCB was applied topically to the skin above the injection site. Fourteen days after the initial injection DNCB was applied to a different skin site and the hypersensitivity response was measured at 24 h on a 6-point scale of erythema.

E. UVB STUDIES

Each guinea pig was exposed to UVB light from a discrete blue point light source for the following times and corresponding intensities of light: 10 s, 10.4 J; 30 s, 31.2 J; 60 s, 62.4 J; 120 s, 124.8 J. Animals were scored for erythema on a 6-point scale of erythema at 4, 24, and 48 h.

F. PHARMACOLOGIC STUDIES

Three topical steroid creams, clobetasol diproprionate (Temovate® 0.05%), fluocinolone acetonide (Lidex® 0.05%), and betamethasone valerate (Valisone® 0.1%), were applied under occlusion to the dorsal skin of guinea pigs. Sites were scored for the occurrence of blanching.

III. RESULTS

A. TISSUE CHARACTERISTICS

Hairless guinea pigs show a significantly greater epidermal thickness compared with albino guinea pigs (Figure 1 and Table 1).

A comparison of the skin proliferation kinetics of hairless and albino guinea pigs is shown in Table 1. The hairless guinea pigs show an increased epidermal transit time and a more rapid turnover of stratum corneum (Table 1).

In vitro percutaneous absorption of [³H]hydrocortisone showed a significant difference between depilated and nondepilated albino guinea pigs (Figure 2,

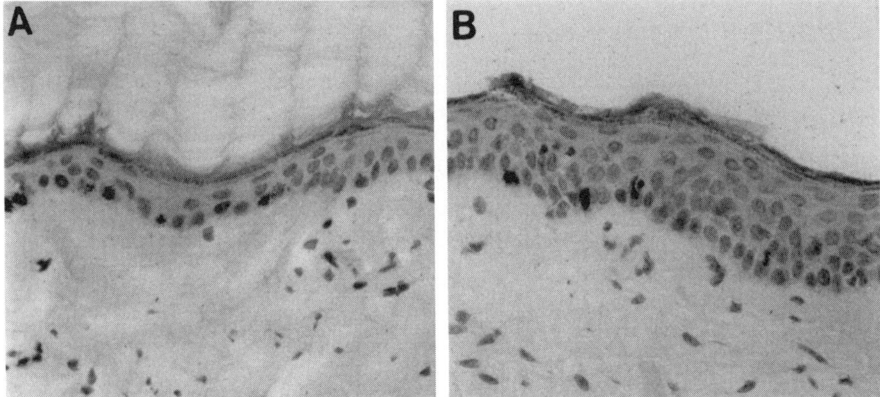

FIGURE 1   Histology of guinea pig skin. (A) Albino, (B) hairless (400×). Dark grains are epidermal cells incorporating tritiated thymidine (1 h labeling).

TABLE 1
Tissue Characteristics

	Epidermal thickness (μm)	Epidermal turnover (days)	Stratum corneum turnover (days)
Hairless	56 ± 14[a]	5	12.3 ± 0.9[a]
Albino	15 ± 2	7	16 ± 0.0

[a] Significantly different $p < 0.005$, t test.

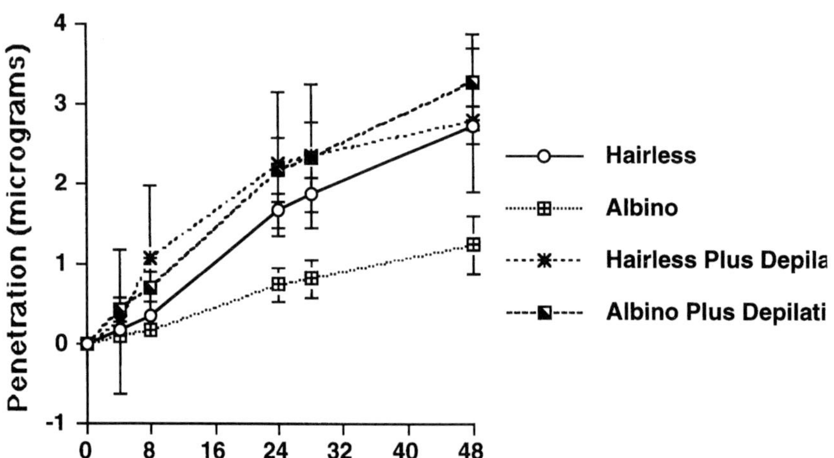

FIGURE 2   Percutaneous absorption of hydrocortisone.

TABLE 2
Erythema Induction by Irritants[a]

	Hairless			Albino		
	Day 1	Day 4	Day 8	Day 1	Day 4	Day 8
Salicylic acid	0	0	0	0	0	0.33 ± 0.47
Formaldehyde	0	0	0	0	0.33 ± 0.75	4 ± 1.41
Benzalkonium Chloride	0	1 ± 0.82	1.33 ± 0.47	2.67 ± 0.75	6[b]	6
Lactic acid	0	2 ± 0.82[b]	0	2 ± 2.2	3.8 ± 0.37[b]	2.67 ± 1.5
Water	0	0	0	0	0	0

[a] Erythema scores based on 0 to 6 scale (see text).

[b] Treatment stopped on this day due to severity of response in albino.

$p < 0.03$, t test, $n = 6$). Nondepilated albinos showed less penetration than either hairless group, particularly as the study progressed. However, depilated albino guinea pigs showed a similar penetration, as did either group of hairless guinea pigs ($p > 0.06$, t test, $n = 6$). Depilation did not significantly affect the results of either hairless group ($p > 0.3$, t test, $n = 6$).

B. IRRITANT STUDIES

Hairless guinea pigs exhibited significantly less irritation to the chemical test agents than did albinos (Table 2). Albinos showed greater maximum irritation than did hairless in all cases. Onset of irritation was slower in the hairless, as was observed with 5% benzalkonium chloride. Furthermore, hairless guinea pigs showed a rapid reversal of the irritation caused by lactic acid when treatment was stopped on day 4. In an experiment where hairless guinea pigs were treated with Nair® an enhanced response to the irritants was observed; however, the irritant response was still less than in albinos (data not shown).

Studies with varying concentrations of retinoic acid showed a similar profile with the other irritants. Once again, albinos showed a greater sensitivity to the irritant effects of retinoic acid as well as a greater overall effect (Table 3). Furthermore, albinos show a faster rate of irritation, particularly at the lower concentrations of retinoic acid.

C. DELAYED CONTACT HYPERSENSITIVITY

Erythema caused by DNCB rechallenge occcurred in both hairless and albino guinea pigs. At 24 h both strains showed a moderate erythema of 3.2 ± 1.0 for albinos and a 3.5 ± 0.5 for hairless.

TABLE 3
Irritation to Retinoic Acid[a]

Retinoic acid	Hairless		Albino	
	Day 4	Day 8	Day 4	Day 8
0.1%	0.92 ± 0.2[b]	2.7 ± 0.3[b]	3.0 ± 0.8	4
0.05%	0.6 ± 0.5[b]	3[b]	2.5 ± 0.6	4
0.005%	0.2 ± 0.5[b]	0.9 ± 0.2[b]	1	2.4 ± 0.5
0.0005%	0.2 ± 0.5[b]	0.6 ± 0.5	1.2 ± 0.3	1.5 ± 0.6

[a] Irritation based on 0 to 6 scale.

[b] Significantly different from albino, $p < 0.05$, t test.

TABLE 4
Erythema Response to UVB[a]

UVB exposure	Hairless			Albino		
	4 h	24 h	48 h	4 h	24 h	48 h
10 s	0	0	0	0.2 ± 0.4	0	0
30 s	0.25 ± 0.43	0.25 ± 0.43	0	1.2 ± 0.98	1.2 ± 1.17	0.4 ± 0.49
60 s	1.75 ± 0.43	1.5 ± 0.5	1.25 ± 1.09	2	2.4 ± 0.49	1 ± 0.63
120 s	2	2	1.5 ± 0.5	2	2.6 ± 0.49	2 ± 0.89

[a] Erythema based on 0 to 6 scale.

D. RESPONSE TO UVB

Hairless guinea pigs exhibited a lower overall response to blue point UVB radiation (Table 4). The hairless group did not show a significant erythema response to UVB until 60 s of exposure, whereas in the albino group approximately 60% of the animals responded to a UVB dose of 30 s.

E. PHARMACOLOGIC RESPONSES

Both hairless and albino guinea pigs responded with equal sensitivity to high potency occluded steroids Temovate® and Lidex® (Figure 3). However, albinos showed a higher sensitivity to the low potency steroid Valisone®.

IV. DISCUSSION

Several factors determine the suitability of a particular animal as a model for experimentation. One factor is the ease of usage for a particular animal. The

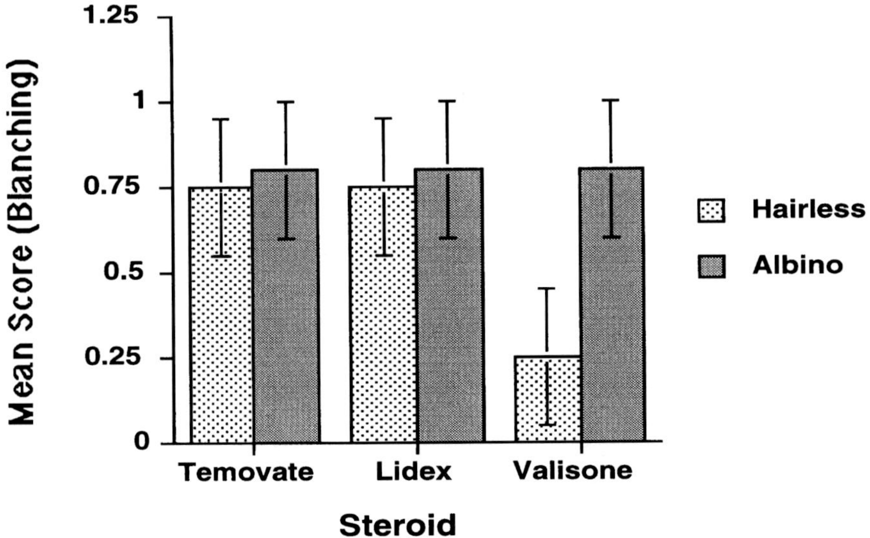

FIGURE 3 Blanching to topical steroid.

hairless guinea pig obviously negates the need of using messy and perhaps irritating depilatory creams and shaving to remove the abundance of hair seen on albino guinea pigs. Therefore, the hairless appears to be ideal for studying topical pharmacologic agents. However, other factors determine the desirability of a model. These include tissue characteristics, sensitivity to chemical agents, and availability of the animals. All of these factors have been considered in this study.

Obvious differences exist between albino and hairless guinea pig skin. We have shown that the epidermis is significantly thicker in the hairless than in the albino. In fact the hairless epidermal thickness more closely resembles the human than does the albino (Richard Guy, personal communication). Along with the greater epidermal thickness is a faster turnover time for both the epidermis and stratum corneum of the hairless. In this case, however, the albino's turnover times are more closely matched to humans than is the skin of the rapidly turning over hairless.[4]

Percutaneous absorption of hydrocortisone *in vitro* does not differ between the depilated albino and the hairless guinea pig. One possible explanation for the difference between the percutaneous absorption of the clipped albino and the depilated albino is the existence of a stubble of hair remaining on the clipped albino. This hair may act as a nondiffusible reservoir for whatever compound is being applied to the guinea pig. This difference is not important to this study as all albinos were depilated before the application of the test compounds.

Overall, the skin of the hairless guinea pig was less sensitive to the test irritants both in intensity of effect and onset of erythema. One speculative explanation for this observation may lay in the epidermis. The epidermis contains enzymes such as arylhydrocarbon hydroxylase, which can biotransform xenobiotics.[4] Possibly in the hairless guinea pig, which has a thicker epidermis, greater biotransformation of the irritants occurs, which may lead to a lower noxious effect of the irritants. Clearly further investigations into the possible differences between biotransformation capabilities of albino and hairless guinea pigs should be studied.

Another inducer of erythema is UV radiation. The skin of the hairless guinea pig was less sensitive than the albino to the effects of UV. Once again differences in skin structure may account partially for the difference in sensitivity to UV. Anderson and Parrish[5] studied the effects of tissue optics on UV exposure. Both hairless and albino guinea pigs are considered to be amelanotic, as both strains have red eyes. The minimal amount of hair that is found on the hairless is white. Another factor which affects tissue optics is epidermal thickness, which may indicate why the hairless is less susceptible to UV.

An interesting finding that we have observed with the hairless guinea pig is the rapid reversal of the erythemogenic effects of irritants such as lactic acid and UV. The rate of reversal is significantly faster than the albino. Given the accelerated rate of turnover for both stratum corneum and epidermis in the hairless guinea pig it is not surprising that the hairless is capable of a rapid recovery.

Although many of the currently available hairless animals are athymic or hypothymic the hairless guinea pig is euthymic. Immunologic function, when based on delayed contact hypersensitivity, does not vary between the strains of guinea pigs. Our findings are in agreement with studies performed for Charles River Laboratories, which concluded that hairless guinea pigs did not have deficient immune responses to contact hypersensitizers (personal communication).

Tests with three classes of topical corticosteroids underscore the differences in pharmacologic sensitivity seen between hairless and albino guinea pigs. Although both hairless and albinos exhibited similar responses to the high potency steroids Temovate and Lidex, the hairless was significantly less sensitive to the lower potency Valisone. This lower sensitivity cannot be explained by a difference in percutaneous absorption, as both hairless and albino show similar penetration profiles for steroids (see above).

An important criterion for using a particular animal model is availability of that species. The commercial availability of the hairless guinea pig has been limited as a communicable disease (pink eye) has necessitated its temporary removal from laboratory use. This factor, if not remedied, will seriously prevent any further use of this animal.

In summary, a variety of differences exist between the hairless and albino guinea pig. Of primary importance to chemical and drug testing is the overall lower sensitivity of the hairless to irritants and pharmacologic agents such as steroids. This is not to say that the hairless is not without some advantages. Studies with agents such as 2,2′-dichlorodiethyl sulfide (sulfur mustard) show that hairless guinea pigs are a superior model to the albino for inducing microvesicle formation, which is similar to lesions observed in humans (Yourick et al.,[6] and personal communication). Given the physical and histological similarities of hairless guinea pig to human skin, this strain of animal may be better suited for the physical-based dermatology models such as dermabrasion and dermal wound healing. Furthermore, since the hairless is euthymic it may be a useful model for opportunistic infection models. Overall, the usefulness of the hairless guinea pig in dermatological research may be restricted to a limited number of applications.

References

1. **Draize, J.H., Woodward, G., and Calvery, H.O.,** Methods for the study of irritation and toxicity of substances applied topically to the skin and mucous membranes, *J. Exp. Pharmacol. Ther.*, 82, 377, 1944.
2. **Anderson, K.E. and Maibach, H.I.,** Guinea pig sensitization assays, *Curr. Prob. Dermatol.*, 14, 263, 1985.
3. **Johnson, A.W. and Goodwin, B.F.J.,** The Draize test and modifications, *Curr. Prob. Dermatol.*, 14, 31, 1985.
4. **Emmett, E.,** Toxic responses of the skin, in *Casarett and Doull's Toxicology: The Basic Science of Poisons*, Amdur, M.A., Doull, J., Klaasen, C.D., Eds., McGraw-Hill, New York, 1991, 463.
5. **Anderson, R.R. and Parrish, J.A.,** The optics of human skin, *J. Invest. Dermatol.*, 77, 13, 1981.
6. **Yourick, J.J., Dawson, J.S., Benton, C.D., Craig, M.E., and Mitcheltree, L.W.,** Pathogenesis of 2,2′-dichlorodiethyl sulfide in hairless guinea pigs, *Toxicology*, 84, 185, 1993.

2 The Use of Epidermal and Dermal T-Lymphocyte Clones in Studies on the Pathophysiology of Bullous Drug Eruptions

Michael Hertl, Doris Niederau, and Hans F. Merk

TABLE OF CONTENTS

I. INTRODUCTION

Drug-induced cutaneous hypersensitivity reactions comprise a variety of distinct clinical entities.[11] While immediate type hypersensitivity reactions, such as urticarial exanthems and anaphylaxis, are mediated by drug-specific IgE antibodies, the pathogenesis of delayed type hypersensitivity reactions, such as morbilliform or bullous eruptions, has not been fully elucidated. Sulfonamides, β-lactam antibiotics, antiepileptics (particularly phenytoin), and nonsteroidal antiinflammatory drugs are the major causes of drug-induced toxic epidermal necrolysis (TEN) and Stevens–Johnson syndrome (SJS).[11,17] There is *in vivo* and *in vitro* evidence that T-lymphocytes are involved in the pathogenesis of drug-induced cutaneous reactions in that peripheral blood mononuclear cells (PBMC) from patients with drug-induced immediate type (urticaria) and delayed hypersensitivity reactions (morbilliform and bullous exanthems) can be frequently stimulated *in vitro* with the causative drugs.[5,6,19,21] In addition, several histological studies have demonstrated that the epidermal inflammatory infiltrate in drug-induced delayed cutaneous hypersensitivity reactions is primarily composed of $CD8^+$ T-lymphocytes.[4,7,13]

Since the role of the T cell subsets associated with drug-induced bullous cutaneous reactions has not been characterized yet, we wanted to study the phenotypic and functional characteristics of epidermal and dermal T-lymphocytes from drug-induced bullous and maculopapular cutaneous reactions and epicutaneous patch test reactions to the causative drugs.

II. MATERIALS AND METHODS

A. PATIENTS

Five β-lactam allergic patients were studied, three of whom developed exanthems resembling moderate toxic epidermal necrolysis with fever, malaise, and progressive blistering 5 to 7 days after treatment with ampicillin (AMPI) or benzylpenicillin (BPO). Two patients developed a maculopapular exanthem on treatment with BPO and three patients exhibited multiforme-like exanthems on treatment with sulfamethoxazole (SMX), carbamazepine, or acetaminophen. All patients had positive epicutaneous reactions following topical application of the causative drug. Skin biopsies were taken after informed consent at the time of onset of the allergic skin reaction or from epicutaneous patch test reactions.

B. CULTURE MEDIA AND STORAGE OF PERIPHERAL BLOOD MONONUCLEAR CELLS

The culture medium uniformly used for the generation of T cell lines and clones and for T cell proliferation assays consisted of RPMI 1640 + 10%

pooled human serum (PHS) with 20 m*M* L-glutamine without antibiotics. The medium used for the expansion of T cell clones and the generation of B-lymphoblastoid cell lines (B-LCL) consisted of RPMI 1640 + 10% fetal calf serum (FCS) with 20 m*M* L-glutamine, 100 U/ml penicillin, and 100 μg/ml streptomycin. Peripheral blood mononuclear cells (PBMC) were purified from venous heparinized blood by gradient centrifugation (Ficoll-Hypaque). If not used immediately, PBMC were kept frozen in liquid N_2 in aliquots of 6 to 10×10^6 cells in a medium consisting of FCS and dimethyl sulfoxide (DMSO) at a ratio of 10:1. The viability of PBMC after thawing ranged from 95 to 100%. T cell clones and B-LCL were frozen in aliquots of 6 to 10×10^6 cells under identical conditions.

C. ISOLATION OF EPIDERMAL AND DERMAL T-LYMPHOCYTES FROM CUTANEOUS DRUG ERUPTIONS

Punch biopsies of 4 to 6 mm or spindle-shaped biopsies were taken from early lesions of drug exanthems or patch test reactions (which were disinfected by topical application of 96% ethanol for 5 min) and were kept overnight in a 0.25% trypsin solution at 4°C (Figure 1).[7,10] The epidermis was then separated from the dermis using forceps, and a single epidermal cell suspension was obtained by mechanical disaggregation. Epidermal cells were cultured with 2×10^6 autologous X-irradiated (5000 R) peripheral blood mononuclear cells (PBMC) and appropriate concentrations of the drug (Table 1) in culture medium. In some cases (if the causative drug was not clearly identified), epidermal T cells were expanded in the presence of autologous, X-irradiated PBMC and 1% phytohemagglutinin (PHA) with 50 U/ml interleukin-2 (IL-2). After 24 h, recombinant human IL-2 (rhIL-2) was added at a final concentration of 50 U/ml to primary cultures stimulated with drugs. In most instances, epidermal T cell blasts were visible in these cultures after 7 to 10 days and could be expanded to the magnitude of 1 to 4×10^6 cells within 3 weeks. In some instances, the dermal portion of the biopsy was disaggregated using forceps and tissue fragments were cultured — analogous to epidermal cells — in the presence of X-irradiated (5000 R) autologous PBMC and the particular drug.[16] Again, rhIL-2 was added at a final concentration of 50 U/ml after 24 h. The expansion and maintenance of dermal T-cells were performed as described for the epidermal T cells.

D. CLONING OF LESIONAL T CELLS BY LIMITING DILUTION

T cell blasts from primary cultures with epidermal or dermal T cells were washed three times in RPMI 1640 to remove unbound IL-2. T cells were cloned by limiting dilution to 0.6 to 1 cell/20-μl well in Terasaki trays in the

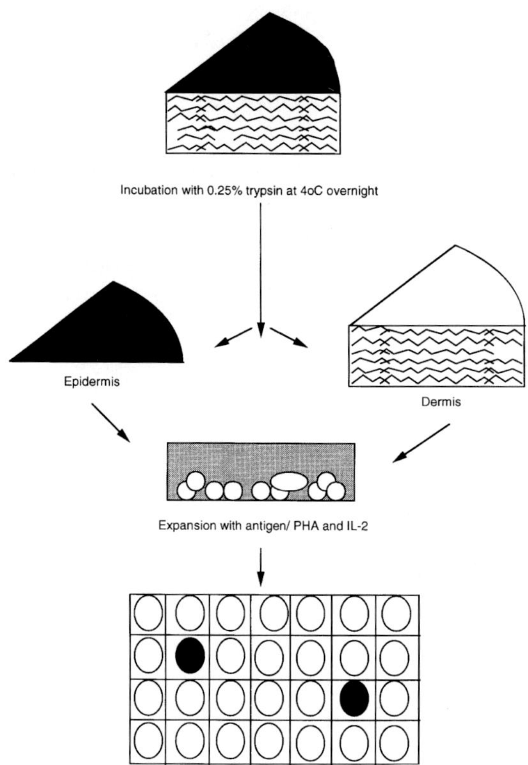

FIGURE 1   Isolation of epidermal and dermal T-lymphocytes from drug eruptions.

presence of 3×10^4 autologous, X-irradiated PBMC as antigen-presenting cells (APC) and the particular drug. Again, rhIL-2 (50 U/ml) was added after 24 h. There were several advantages in using Terasaki trays for the cloning procedure. First, smaller amounts of autologous PBMC as feeder cells were required. Since peripheral blood could be obtained from these patients only a few times in most instances, the supply in autologous PBMC was the limiting factor for these studies. Second, the milieu provided by the small volume of these cultures seemed to favor the outgrowth of T cell clones. In our hands, the screening procedure for growing T cell clones was easier in Terasaki trays than in microtiter plates. After 1 to 2 weeks, growing T cell clones were transferred to flat-bottom microtiter culture plates and were expanded in culture medium (RPMI + 10% PHS) supplemented with 50 U/ml rIL-2. Once the T cell clones became unresponsive to IL-2, they were restimulated at 6×10^5 with 1% PHA or anti-CD3 monoclonal antibodies (1 ng/ml) and 2×10^6 X-irradiated PBMC as APC. Culture media were changed on day 1 to 2 after stimulation to reduce the amount of unbound PHA, and T cells were cultured in medium (RPMI + 10% FCS) supplemented with 50 U/ml rIL-2 at a density of 1 to 2×10^6 cells/ml.

TABLE 1
**Concentrations of Drugs Used for *In Vitro*
Proliferation Assays with T-Lymphocytes**

Drug	Concentration (µg/ml)
Acetaminophen	2–100
Acetylsalicylic acid	10–50
Ampicillin	50–500
Benzlpenicillin	500–2000
Carbamazepine	10–50
Cephalosporin	10–100
Ciprofloxacin	5–100
Ibuprofen	5–50
Indomethacin	5–50
Metamizol	20–100
Phenazon	20–100
Phenytoin	5–100
Propyphenazon	20–10
Sulfamethoxazole	100–500
Tetracyclin	2–30
Trimethoprim	10–250

E. DERIVATION OF DRUG-SPECIFIC T CELL CLONES FROM THE PERIPHERAL BLOOD

PBMC from patients with β-lactam-induced maculopapular or bullous exanthems were plated at 3×10^5 cells/200-µl flat bottom microtiter culture plates in culture medium (RPMI + 10% PHS) in the presence of the particular causative drug at different concentrations (Table 1).[7] A parallel plate (B) with an identical setup was also cultured for 7 days. While drug-specific lymphocyte proliferation was determined on the first plate (A) by [3H]thymidine uptake, proliferating drug-specific T cells were expanded on plate B by adding rIL-2 at a final concentration of 50 U/ml. These proliferating T cell blasts were either restimulated with autologous PBMC as APC and the particular drug or were immediately cloned by limiting dilution. For the restimulation (secondary culture), two parallel plates (with 200-µl microtiter wells) were set up with 4 to 5×10^4 T cells from plate B (primary culture) and 5×10^4 autologous, X-irradiated PBMC in the presence of the particular drug and were incubated for 4 days. The uptake of [3H]thymidine by drug-specific T cells on one plate served as a control for the specificity of the expanded cells on the second plate. T cells from the second or third restimulation were cloned at 0.5 to 1 cells/20-µl culture in Terasaki trays with 3×10^4 X-irradiated autologous PBMC and the drug. Recombinant IL-2 was added to these cultures at a final concentration of

50 U/ml after 24 h. Proliferating T cell clones were expanded with PHA or anti-CD3 as described above.

F.   GENERATION OF B-LYMPOBLASTOID CELL LINES (B-LCL)

PBMC from patients with drug-induced cutaneous reactions were cultured at 2 to 3×10^6/ml in RPMI + 10% FCS supplemented with the supernatant (30%, v:v) from the marmose T cell line B95 as a source of Epstein–Barr virus (EBV). The culture medium also contained 500 ng cyclosporin A (CsA) to prevent the killing of EBV-transformed B-lymphoblastoid cells (B-LCL) by cytotoxic T cells. Fresh medium containing 50 ng CsA was added once the first clusters of outgrowing B-LCL were visible. B-LCL were cultured in medium supplemented with CsA until contaminating T cells were no longer detectable. Autologous B-LCL were a reasonable source of APC for proliferation assays with drug-specific T cell clones in that they provided good antigen-specific stimulation for tetanus toxoid-specific and penicillin-specific T cell clones. B-LCL were X-irradiated (8000 to 10,000 R) prior to coculture with T cells to reduce background proliferation.

G.   PROLIFERATION ASSAYS WITH T CELL CLONES FROM DRUG EXANTHEMS

Samples of 2 to 5×10^4 epidermal or dermal T cells (which were rested in RPMI + 10% FCS + 25 U/ml rIL-2 for 7 to 14 days after stimulation) were cultured with X-irradiated autologous or allogeneic APC in 200-μl flat bottom microtiter plates for 72 h at 37°C in 5% CO_2. The sources of APC were frozen autologous PBMC, which were X-irradiated (5000 R) and added at 5×10^4 or X-irradiated autologous B-LCL (2 to 3×10^4). In proliferation assays with penicillin-specific T cell clones, APC were incubated with 10-fold concentrations of the particular drug (BPO 10,000 U/ml; AMPI 5000 μg/ml) in RPMI 1640 for 1 h at 37°C prior to coculture with T cells. APC were then washed extensively to remove unbound antigen. T cell proliferation was determined by the uptake of [^{3}H]thymidine after 72 h.

H.   DRUG-SPECIFIC T CELL PROLIFERATION IN THE PRESENCE OF DRUG-MODIFIED MURINE LIVER MICROSOMES

Murine liver microsomes were prepared using a standard protocol.[12,15] Female NMRI mice were administered orally 50 mg/kg phenobarbital once daily over 3 days to induce drug-metabolizing cytochrome P-450 enzymes known to metabolize sulfonamides. Livers were removed from mice and minced and microsomes were prepared by differential ultracentrifugation, and were either used immediately or stored in liquid N_2. Protein was determined using bovine serum albumin as a standard. Cytochrome P-450-dependent 7-ethoxyresorufin O-deethylation, 7-pentoxyresorufin O-deethylation, and p-nitrophenol hydroxylation were determined as described elsewhere.[9] Microsomal cytochrome P-450 2B-dependent 7-pentoxyresorufin O-deethylase activity was induced (57.7 ± 2.1 to 498 ± 5.2 mmol resorufin/min/mg protein) whereas the activity of cytochrome P-450 1A-dependent 7-ethoxyresorufin O-deethylase was not significantly altered.[8] These differential isoenzyme activities were paralleled by a predominant staining of microsomal proteins in Western blots with polyclonal antibodies (Poab) against cytochrome P-450 2B isoenzymes and only a weak staining with PoAb against cytochrome P-450 1A isoenzymes.[8] Dermal T cells (2 to 5×10^4) were incubated with 2 to 5×10^4 X-irradiated (5000 R) autologous PBMC as APC in 200-µl flat bottom microtiter plates for 72 h at 37°C in 5% CO_2. After 24 h, cultures with microsomes were spun down and were resuspended in fresh culture medium (RPMI + 10% PHS) to avoid toxic effects of microsomes. T cell proliferation was determined by the uptake of [^{3}H]thymidine.

I.   CYTOKINE PRODUCTION OF EPIDERMAL T CELL CLONES

Samples of 10^6 epidermal or dermal T cell clones from cutaneous drug reactions were cocultured with 5×10^5 autologous, X-irradiated B-LCL and 1% PHA in culture medium (RPMI + 5% FCS) for 1 h at 37°C and 5% CO_2. Afterward, cultures were washed extensively to remove unbound PHA and cells were then cultured in the identical medium without PHA. Supernatants (SN) were collected in the identical medium without PHA. Supernatants (SN) were collected from these cultures after 72 h and were centrifuged to remove cellular components. SN were also collected from cultures with B-LCL and PHA without T cells, which served as controls to determine whether B-LCL also produced the investigated cytokines. Alternatively, 10^6 T cells were stimulated with 10 ng/ml phorbol myristate acetate (PMA) and 0.5 µg/ml ionomycin in RPMI + 5% FCS for 4 h, washed 2 times afterward and cultured in RPM + 5% FCS for an additional 12 h at 37°C. At the end of the incubation period, culture SN were collected and stored at 4°C. We favored the system with PMA and ionomycin, which allowed for maximal T cell stimulation without contaminating APC as a potential source of cytokine production. Interferon-gamma (IFN-γ), interleukin-4 (IL-4), and granulocyte–macrophage colony-stimulating factor (GM-CSF) immunoreactivity in the SN was determined using commercial enzyme-linked immunosorbent assays (ELISA). IL-2 bioactivity was analyzed using the IL-2-sensitive human CD4$^+$ T cell clone MH2. In short, 5×10^3 MH2 T cells (in the resting phase) were incubated with serial dilutions of recombinant human IL-2 (standards) or dilutions of the culture SN (1:4, 1:8, 1:16) in culture medium (RPMI + 5% FCS) for 18 h.

IL-2-dependent T cell proliferation of MH2 was determined after 24 h by the uptake of [^{3}H]thymidine, which was added 6 h before the termination of the assay. Proliferation of MH2 was strictly IL-2 dependent and could be blocked in the presence of anti-IL-2 antibodies.

III. RESULTS

A. PREDOMINANCE OF CD8[+] T CELL CLONES FROM MORBILLIFORM AND BULLOUS DRUG ERUPTIONS

We have isolated exclusively CD8[+] epidermal T cell clones from drug-induced bullous exanthems and patch test reactions.[6] Three epidermal CD8[+] T cell clones from β-lactam-induced bullous exanthems proliferated to the particular causative drug in an MHC-restricted fashion (Figure 2) and produced TH1 pattern cytokines after activation (Figure 3).[6] Identical observations were made with peripheral drug-specific T cell clones from patients with β-lactam-induced morbilliform exanthems that were exclusively CD8[+] and proliferated

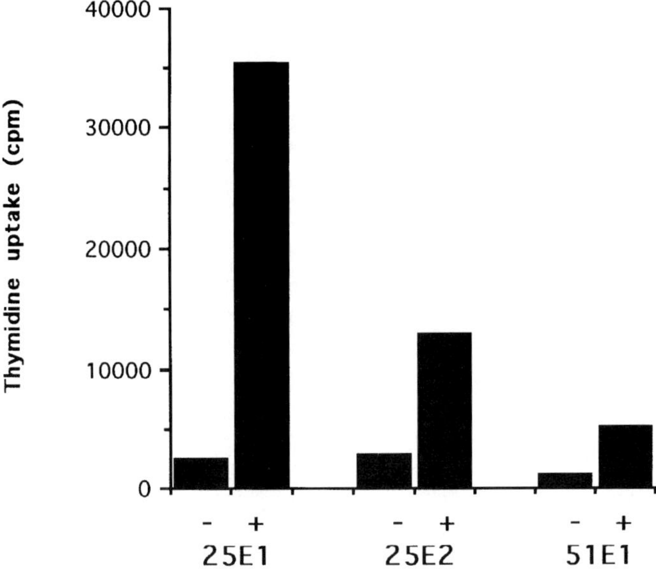

FIGURE 2 Proliferation assay with β-lactam-specific epidermal CD8[+] T cell clones from β-lactam-induced bullous exanthems. T cell clones 25E1, 25E2, and 51E1 were incubated with X-irradiated autologous B-LCL in 200-μl cultures for a period of 72 h. B-LCL were added to cultures untreated (-) or after preincubation with ampicillin (+) for T cell clone 51E1-2 or with BPO (+) for T cell clone 51E1, respectively. T cell proliferation was determined by [^{3}H]thymidine uptake after 72 h, and is expressed as counts per minute (cpm) for triplicate cultures. Standard deviation never exceeded 10% of the mean.

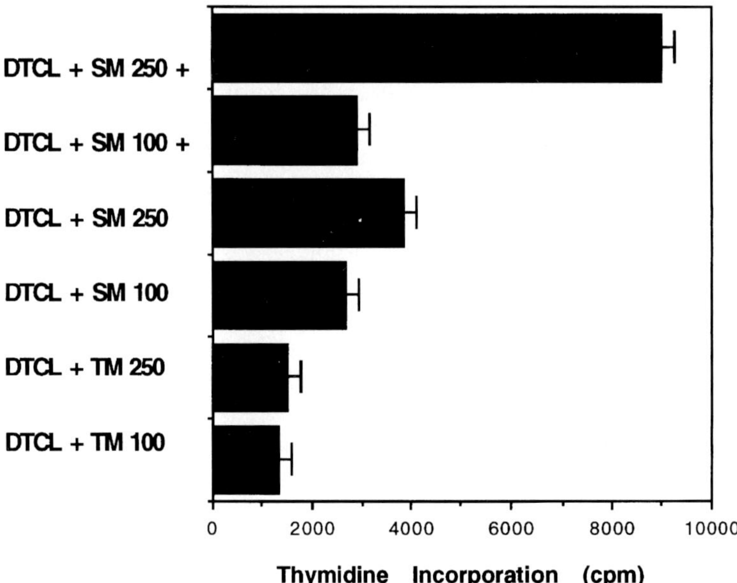

FIGURE 3 Effect of murine microsomes enriched for cytochrome P-450 2B isoenzymes on the proliferative response of a CD8[+] dermal T cell line to sulfamethoxazole. Dermal CD8[+] T cells (DTCL) were incubated with X-irradiated autologous PBMC as APC in 200-μl cultures over 72 h. Identical cocultures were performed in the presence of murine liver microsomes (+). Trimethoprim (TM) and sulfamethoxazole (SM) were used at a final concentration of 100 to 250 μg/ml. T cell proliferation was determined by [^{3}H]thymidine incorporation and is expressed as counts per minute (cpm ± SD).

in response to BPO and also to the penicilloyl moiety (PL), which is the major metabolite of BPO.[7] These clones also secreted Th1-like cytokines. CD8[+] T-lymphocytes have also been identified as effector cells in allergic contact dermatitis[3] and in humans allergic to the hapten poison ivy.[10] All β-lactam-specific epidermal CD8[+] T cell clones produced significant amounts of IFN-γ, which is known to induce up-regulation of ICAM-1 on endothelial cells and keratinocytes.[6,7,14] The up-regulation of ICAM-1 on keratinocytes in cutaneous drug reactions—which can be detected *in situ* by immunohistochemistry[6]—may be the consequence of the release of IFN-γ by infiltrating T cells. ICAM-1, the ligand for LFA-1 on T cells, is known to be a critical adhesion molecule for the homing of T-lymphocytes to the skin.[14]

B. DRUG METABOLISM BY MICROSOMAL ENZYMES AND DRUG-SPECIFIC T CELL ACTIVATION

Most drugs, including sulfonamides, are metabolized to reactive species by cytochrome P-450-dependent enzymes.[1,12,18] Recent studies indicate that some patients with sulfonamide-induced bullous cutaneous eruptions exhibit reduced N-acetylation capacity.[2,20] However, the relation between this defect and the metabolism of sulfonamides is still unclear because sulfonamides are acetylated by the noninducible acetyltransferase isoenzyme (S. Leder, personal communication). Is this, therefore, of interest that patients with an increased cytochrome P-450-dependent metabolism of sulfonamides to hydroxylamine derivatives have an increased risk to develop drug-induced hypersensitivity reactions, suggesting that the relationship between sulfonamide metabolism and hypersensitivity is even more complex. Our model system with lesional sulfamethoxazole (SMX)-specific T cell clones provides the possibility to study this relationship more precisely. Our findings suggest that cytochrome P-450-dependent SMX metabolites are relevant T cell antigens in that a sulfamethoxazole-specific CD8+ dermal T cell clone showed a significantly increased proliferation to SMX in the presence of cytochrome P-450-rich microsomes (Figure 4).[8]

IV. SUMMARY AND CONCLUSIONS

Drug-specific CD8+ T-lymphocytes are present in the peripheral blood and involved skin of patients with drug-induced delayed cutaneous hypersensitivity reactions, such as morbilliform and bullous exanthems. The observed predominant of epidermal CD8+ T-lymphocytes in drug-induced bullous exanthems suggests that this particular T cell subset plays a role in the pathogenesis of drug-induced blister formation. The metabolism of the many drugs, for example, by cytochrome P-450-dependent enzymes, may be critical in the formation of the normal T cell antigens, as suggested in the case

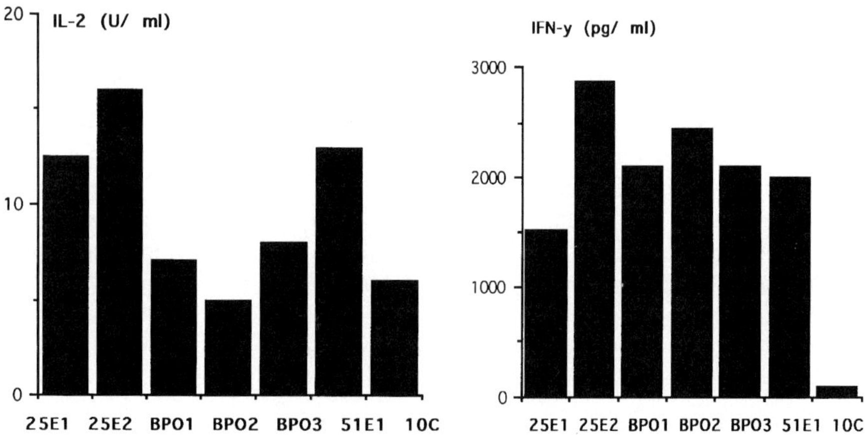

FIGURE 4   Cytokine production of CD8+ epidermal T cell clones. IL-2 bioactivity is expressed as U/ml and immunoreactivity of IFN-γ, as determined by ELISA, is expressed as pg/ml.

of SMX-dependent T cell activation. This model system with epidermal and dermal T cell clones from cutaneous drug eruptions has the potential to provide a better understanding of the functional characteristics of lesional T-lymphocytes and of the molecular events leading to drug-induced T cell activation.

ACKNOWLEDGMENTS

This work has been supported by Grants He 1602/3-1 and He 1602/3-2 from the Deutsche Forschungsgemeinschaft, Bonn, Germany. We thank Dr. Andrea Cavani for critically reading the manuscript.

References

1. **Cribb, A.E. and Spielberg, S.P.,** Sulfamethoxazole is metabolized to the hydroxylamine in humans, *Clin. Pharmacol. Ther.*, 31, 522, 1992.
2. **Dietrich, A., Kawakubo, Y., Rzany, B. et al.,** Patients with severe cutaneous drug reactions (EM, SJS, TEN) have a reduced N-acetylating capacity, *J. Invest. Dermatol.*, 100, 519A, 1993.
3. **Gocinski, B.L. and Tigelaar, R.E.,** Roles of CD4+ and CD8+ T-cells in murine contact sensitivity revealed by in vitro monoclonal antibody depletion, *J. Immunol.*, 144, 4121, 1990.
4. **Heng, M.C.Y. and Allen, S.G.,** Efficacy of cyclophosphamide in toxic epidermal necrolysis, *J. Am. Acad. Dermatol.*, 25, 778, 1991.
5. **Hertl, M., Schneider, R., and Merk, H.F.,** In vitro-lymphocyte responses to beta-lactam antibiotics in penicillin allergy: cross-reactivity between penicillins and cephalosporins, *Dermatosen*, 40, 102, 1992.
6. **Hertl, M., Bohlen, H., Jugert, F., Boecker, C., Knaup, R., and Merk, H.F.,** Predominance of epidermal CD8+ T-lymphocytes with bullous cutaneous reactions caused by beta-lactam antibiotics, *J. Invest. Dermatol.*, 101, 794, 1993.
7. **Hertl, M., Geisel, J., Boecker C., and Merk, H.F.,** Selective generation of CD8+ T-cell clones from the peripheral blood of patients with cutaneous reactions to beta-lactam antibiotics, *Br. J. Dermatol.*, 128, 619, 1993.
8. **Hertl, M., Jugert, F., and Merk, H.F.,** CD8+ dermal T-cells from a sulfamethoxazole-induced bullous exanthem proliferate in response to drug-modified liver microsomes, *Br. J. Dermatol.*, in press.
9. **Jugert, F., Agarwal, R., Kuhn, A., Bickers, D.R., Merk, H.F., and Mukhtar, H.,** Multiple cytochrome P450 isozymes in murine skin: induction of P450 1A, 2B, and 3A by dexamethasone, *J Invest. Dermatol.*, 102, 970, 1994.

10. **Kalish, R.S. and Johnson, K.L.,** Enrichment and function of urushiol-specific T-lymphocytes in lesions of allergic contact dermatitis to urushiol, *J. Immunol.*, 145, 3706, 1990.

11. **Kalish, R.S.,** Drug eruptions: a review of clinical and immunological features, *Adv. Dermatol.*, 6, 221, 1991.

12. **Merk, H.F., Schneider, R., and Scholl, P.,** Lymphocyte stimulation with drug-modified microsomes, in *Toxicological and Immunological Aspects of Drug Metabolism and Environmental Chemicals*, Eastbrook, R.W., Lindlaub, E., Oesch, R., and de Weck, A.L., Eds., Schattauer, Stuttgart, Germany, 1987, 211–220.

13. **Miyauchi, H., Hosokawa, H., Akaeda, T., Iba, H., and Asada, Y.,** T-cell subsets in toxic epidermal necrolysis, *Arch. Dermatol.*, 127, 851, 1991.

14. **Nickoloff, B.J.,** Role of gamma interferon in cutaneous trafficking of lymphocytes with emphasis on molecular and cellular adhesion events, *Arch. Dermatol.*, 124, 1835, 1988.

15. **Niederau, D., Fritzsche, R., and Merk, H.F.,** Lymphocyte transformation test with drug-modified microsomes, *Z. Hautkr.*, 68, 581, 1993.

16. **Reinhold, U., Pawelec, G., Fratila, A., Leippold, S., Bauer, R., and Kreysel, H.W.,** Phenotypic and functional characterization of tumor infiltrating lymphocytes in mycosis fungoides: continuous growth of CD4+ CD45R+ T-cell clones with suppressor-inducer activity, *J. Invest. Dermatol.*, 94, 304, 1990.

17. **Roujeau, J.C.,** The spectrum of Stevens-Johnson syndrome and toxic epidermal necrolysis: a clinical classification, *J. Invest. Dermatol.*, 102, 28s, 1994.

18. **Shear, N.H. and Spielberg, S.P.,** Anticonvulsant hypersensitivity syndrome: in vitro assessment of risk, *J. Clin. Invest.*, 82, 1826, 1988.

19. **Stejskal, V.D.M., Forsbeck, M., and Olin, R.,** Side chain-specific lymphocyte responses in workers with occupational allergy induced by penicillins, *Int. Arch. Allergy Appl. Immunol.*, 82, 461, 1987.

20. **Wolkenstein, P., Carriere, V., Charue, D., Revuz, J., Roujeau, J.C., Beaune, Ph., and Bagot, M.,** Slow acetylator genotype as a metabolic predisposition to sulfonamide-induced toxic epidermal necrolysis, *J. Invest. Dermatol.*, 103, 423A, 1994.

20. **Zakrzewska, J.M. and Ivnyi, L.,** In vitro lymphocyte proliferation by carbamazepine, carbamazepine-10, 11-epoxide, amd oxcarbamazepine in the diagnosis of drug-induced hypersensitivity, *J. Allergy Clin. Immunol.*, 82, 110, 1988.

3 *In Vivo* Model of Cutaneous Inflammation Induced by Subsets of T Helper Cells in Mice

Kai M. Müller and Conrad Hauser

TABLE OF CONTENTS

1.   INTRODUCTION

The classical delayed-type hypersensitivity (DTH)* reaction has long been known to be T cell dependent. These reactions are usually elicited in immunized humans and laboratory animals by the epicutaneous application of haptens or the intracutaneous injection of proteins or microorganisms. T-lymphocytes are involved in many forms of skin inflammation. They may indirectly be responsible for inflammatory reactions such as IgE-dependent immediate-type hypersensitivity reactions by inducing the switch toward IgE production

---

* Abbreviations: Con A, concanavalin A; DTH, delayed-type hypersensitivity; DTH1, inflammatory reaction mediated by Th1 cells; DTH2, inflammatory reaction mediated by Th2 cells; mAb, monoclonal antibody; MHC, major histocompatibility complex; sIL-4R, soluble IL-4 receptor, TCR, T cell receptor, Th, T helper cell.

in B cells. T cells may, however, more directly induce skin inflammation by the release of proinflammatory cytokines.

A complex series of events is thought to lead to the recruitment of specific T cells to the site of antigen deposition followed by their activation through interaction with local antigen-presenting cells. Activation is followed by the release of a number of cytokines from the T cells. These cytokines are considered to be responsible for the subsequent inflammatory phenomena. Although the specificity of the clonally distributed T cell receptor (TCR) plays an important role for antigen recognition in the context of major histocompatibility complex gene products, the sets of cytokines released in response to antigen triggering by the TCR, i.e., the spectrum of cytokines released by activated T cells, is restricted and not directly related to the structure of the TCR.

The function of T helper (Th) cells is closely related to the cytokines they secrete upon activation. Whereas mature, nonprimed resting Th cells activated *in vitro* produce essentially interleukin-2 (IL-2), which serves as an autocrine growth factor, previously activated ("antigen-experienced") CD4[+] T cells may produce a vast array of cytokines when restimulated and thereby become capable of subserving various effector functions. In 1986, Mosmann and co-workers became aware of the fact that many long-term cultured Th cell clones produced a restricted pattern of cytokines upon reactivation, and could accordingly be classified into two distinct subsets.[1] The first group comprised clones preferentially producing IL-2 and interferon-gamma (IFN-γ) and was termed Th1, while the other subset, termed Th2, consisted of clones that produced large amounts of IL-4 and IL-5, but no detectable IL-2 or IFN-γ.[1] Extension of the analysis to further cytokines confirmed that the two subsets produced cytokines unique for each subset but shared the release of others (Table 1).[2-4] The exclusive cytokine profiles of Th1 and Th2 cells represent extremes of a spectrum of possible cytokine patterns. This became apparent when the existence of so-called Th0 cell clones that have an intermediate cytokine pattern comprising both Th1- and Th2-specific cytokines was reported.[5] The unrestricted cytokine pattern of Th0 cells closely resembled that of polyclonal CD4[+] T cell populations that were restimulated *in vitro* after primary activation. The development of Th cell subsets appears to occur from a common precursor cell[6] and to be driven by different factors: Th1 cells are induced by the combined action of the cytokines IL-2, IFN-γ, and IL-12, while prolonged TCR occupancy in combination with IL-4 promotes the development of the Th2 phenotype.[7-9]

Because of the complexity of the biological events that include recruitment of the T cells into the skin prior to local T cell cytokine release, *in vitro* studies with various agents that block the effect of cytokines and adhesion molecules do not allow us to distinguish between the actual contribution of T cells and preceding events that are required for T cell activation in the skin. We therefore developed an *in vivo* model of T cell-induced skin inflammation that does not require previous processes and that is independent of the specificity of the T cell receptor (TCR). CD4[+] T cells with a distinct cytokine pattern were pulsed with activating antibodies *in vitro* and subsequently injected into the footpads and/or ears of naive recipient mice. Inflammation elicited by these

TABLE 1
Cytokine Profiles of Murine Th1 and Th2 Cells

Cytokine	Production by	
	Th1	Th2
IL-2	+++	—
IFN-γ	+++	—
TNF-β	+++	—
IL-3	++	++
GM-CSF	++	++
TNF-α	++	+
IL-4	—	+++
IL-5	—	+++
IL-6	—	++
IL-9		+++
IL-10	—	+++
IL-13	—	+++

means was monitored by various ways. This system allowed us to define some cytokines and adhesion molecules that are involved in different forms of T cell-mediated skin inflammation *in vivo*. Here, we discuss the various effector functions of Th1 and Th2 cell subsets and provide a detailed description of local tissue inflammation reactions that can be elicited by activated Th cell subsets and depend on different cytokine profiles released by these cells.

II. MATERIALS AND METHODS

CD4[+] T cells with a Th1- or a Th2-like phenotype were obtained as described,[10,11] by activation of CD4[+] T cells in 2 to 3-day cultures with 3000-rad-irradiated syngeneic splenocytes in the presence of concanavalin A (Con A; 2 µg/ml), of the activating monoclonal antibody (mAb) 145-2C11 to CD3ε (1 µg/ml),[12] or of *Staphylococcus aureus* enterotoxin B (10 µg/ml), followed by expansion for 10 to 11 days in medium containing 50 U/ml recombinant human IL-2, in the presence (Th2) or the absence (Th1) of Con A (1 µg/ml) or mAb 145-2C11 (1 µg/ml).

For the assessment of the capacity of these polyclonal Th cell subsets to induce inflammatory reactions *in vivo*, viable Th1 and Th2 cells were incubated on ice with mAb 145-2C11 (1 µg/ml) or the mAb indicated, extensively washed in phosphate-buffered saline (PBS) and injected sc into the hind footpads of healthy nonimmunized recipient animals as described (usually 2.5×10^5 cells in 30 µl PBS).[13-15] The thickness of each paw was measured immediately before and at various times after injection. Swelling was defined as the difference between paw thickness after injection and thickness before injection. Nonpulsed cells were used as negative controls. Standard histological examinations were carried out blinded as described.[14,15]

Cytokine bioactivities were determined in the supernatants of Th1 and Th2 cells restimulated as described.[10,11] Tumor necrosis factor (TNF) was determined by its inhibitory action on the growth of WEHI 164 subclone 13 fibrosarcoma cells.[15] IL-2 and IL-4 were tested in CTLL-2 or CT.4S growth assays and IL-5 and IFN-γ were determined by enzyme-linked immunosorbent assay (ELISA) as described.[10,11,16]

III. RESULTS

The occurrence of Th1- and Th2-like cells has been found to be associated with various appropriate and inappropriate immune responses *in vivo*.[7,17] Th1 cells are known to promote macrophage activation through the release of IFN-γ, and to mediate both TNF-dependent and -independent cytotoxicity;[18,19] these functional activities could be inhibited by Th2 cell cytokines. Most Th1 clones were unable to induce immunoglobulin production by B cells,[20] while most Th2 clones induced IL-4 dependent IgG_1 and IgE synthesis in B cells.[20-22] Our experiments with polyclonal *in vitro*-derived Th cells confirmed the capacity of Th2 cells to induce IgE (and also IgG_1 and IgG_{2A}) synthesis in B cells *in vitro,* whereas Th1 cells failed to provide any B cell help at all.[11]

There is evidence that acute graft-versus-host reactions are mediated by Th1 cells[23,24] and are inhibited by cells with a Th2 phenotype.[25] Conversely, chronic graft-versus-host reactions as well as some autoimmune diseases have been thought to be mediated by Th2-like cells.[26-29] The high titers of virus-neutralizing IgG_{2A} antibodies observed in acute viral infections might result from the activation of Th1-like cells.[30] High IgE levels are commonly observed in parasitic diseases such as infestation with *Nippostrongylus brasiliensis*, and were shown to be decreased by the administration of anti-IL-4, as well as by the injection of IFN-γ, the natural antagonist of IL-4 with regard to IgE synthesis.[31] The best documented example of *in vivo* development of Th1 and Th2 cells and their association with different immune responses has been provided by murine leishmaniasis. Mouse strains resistant to *Leishmania major*, such as C57BL/6 (H-2^b), mounted responses directed by IL-12 and the development of IFN-γ-producing Th1-like cells that induced macrophage activation to a microbiocidal state and led to elimination of the parasites.[32] In contrast, strains susceptible to infection with *Leishmania*, such as BALB/c (H-2^d), mounted a deleterious Th2 response. Neutralization of IL-4 or administration of recombinant IL-12 induced a curative Th1 development in these infected mice.[33] The relationship between the major histocompatibility complex (MHC) haplotype of these inbred mouse strains and their susceptibility to infection with *Leishmania* has not yet been established. In view of these results, the induction of Th1 cells by infections and parasitic infestations has generally been associated with protective immunity, whereas the induction of immune responses with predominant Th2 cells correlated with disease progression or persistence in most of the animal models.[34] Although a clear-cut difference between Th1 and Th2, as observed in mice, has not yet been demonstrated in humans, there is growing evidence that in humans, too, Th2 cytokine patterns may be involved in a series of diseases. Among these, most interest has focused on the lepromatous form of leprosy (in contrast to the tuberculoid form that is dominated by Th1 responses),[35,36] on the development of HIV infection toward the pathophysiology of AIDS,[37] and on the clinical manifestations of atopic dermatitis.[17]

In 1987, Cher and Mosmann demonstrated that Th1 clones were capable of mediating MHC-restricted, antigen-dependent DTH responses when injected into immunologically naive recipient mice (DTH1).[38] In contrast, none of five Th2 clones examined elicited significant tissue reactions at the time of DTH assessment (24 to 72 h). Similar results were subsequently obtained by several other investigators,[39-41] although not all Th1 clones tested induced tissue inflammation.[41] Testing polyclonal Th1 cell populations derived following our protocol, we could subsequently demonstrate that inflammatory tissue reactions could also be observed when such polyclonal Th1 cells were pulsed with non-MHC-restricted activators such as mAb to the TCR/CD3 complex or Thy-1.2[14] (Figure 1).

Similarly to what had been previously described with antigen-specific clones, the DTH1 responses had characteristic delayed kinetics: swelling was detectable at 6 to 9 h, peaked at 24 to 48 h after injection of the cells, and usually lasted for more than 5 days. When the lesions were examined by classical histology, a strong polymorphonuclear infiltrate was seen at 6 h, whereas at 24 to 48 h mononuclear cells prevailed in the vicinity of the injection point.

The inflammatory reaction mediated by Th1 cells was shown to be partially dependent upon IFN-γ[40] and also upon TNF,[14] two cytokines known to be produced in large amounts by Th1 cells. These two cytokines collaborate in eliciting DTH1, since ip pretreatment of recipient animals with a combination of blocking antibodies to IFN-γ and to TNF totally abrogated footpad swelling induced by Th1 cells.[41a] This is consistent with the observation of preferential expression of mRNA for the two Th1-typical cytokines IL-2 and IFN-γ in classical DTH reactions to tuberculin in the human[42] and with the substantial release of TNF from polyclonal Th1 cells restimulated *in vitro*.[15]

Although Cher and Mosmann originally reported that Th2 clones did not mediate DTH-like inflammatory tissue reactions upon local adoptive transfer into naive recipient mice,[38] this study explored only relatively late (24 to 72 h) reactions. By local injection of anti-CD3ε-pulsed polyclonal Th cells into the footpads of recipient mice and subsequent monitoring of paw thickness, we have shown that Th2 cells are indeed capable of inducing inflammatory tissue reactions *in vivo*.[14] This novel type of reaction was termed DTH2 and was shown to be linked to the activating effect of TCR occupancy, since it could also be induced by pulsing Th cells with activating mAb to pan-$\alpha\beta$ TCR or to Thy-1.2, but not by a nonactivating anti-Thy-1.2 mAb or by anti-CD4 (Figure 1). None of these antibodies induced *per se* cytokine release from the pulsed cells in the absence of cross-linking, as demonstrated by subsequent culture *in vitro* and

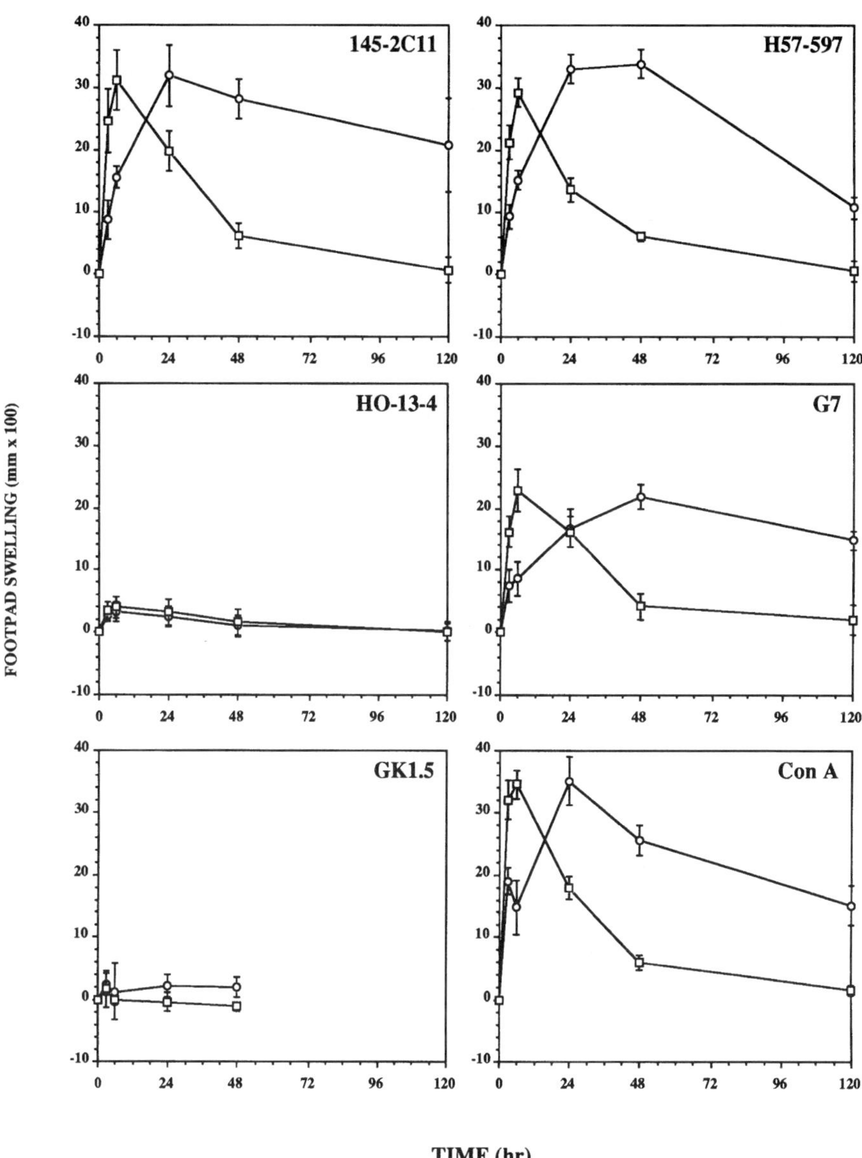

FIGURE 1 *In vitro* derived polyclonal Th1 (open circles) and Th2 (open squares) cells were pulsed for 1 h with the antibodies indicated (1 µg/ml) and subsequently injected into the hind paws of naive recipient mice. 145-2C11, mitogenic anti-CD3ε; H57-597, mitogenic anti-pan-αβ-TCR; HO-13-4, nonmitogenic anti-Thy-1.2; G7, mitogenic anti-Thy-1.2; GK1.5, nonmitogenic anti-CD4; Con A, concanavalin A, a lectin mitogenic to T cells. Data are shown as mean ±SD, $n = 5$.

analysis of the supernatants. Instead, we speculate that *in vivo* cross-linking of these mAb by Fc receptors on bystander cells induced Th cell activation and local cytokine release. The mitogenic lectin Con A was also capable of inducing Th1- and Th2-dependent tissue swelling reactions with their typical kinetics (Figure 1). Moreover, it is not plausible that the pulsed Th cells injected in our system merely served as antibody (or Con A) carriers, since heat treatment of the cells (55°C for 30 min) or their prolonged culture in cytotoxic ethanol concentrations (70% for 30 min), followed by pulsing with mAb 145-2C11, extensive washing, and injection resulted in strongly decreased swelling responses (K.M. Müller and C.Hauser, unpublished). This interpretation is corroborated by the finding that transfer of Th1 or Th2 cells into congenic *nu/nu* mice resulted in swelling responses that were only slightly, if at all, impaired (Figure 2). It furthermore appeared that the inflammatory reactions to Th1 as well as to Th2 cells were dependent on the secretion of cytokines by these cells, since both the DTH1 and DTH2 reaction could be largely inhibited by an overnight culture of the donor cells with cyclosporin A (10 µg/ml) followed by extensive washing prior to injection.[14] Experiments with antigen-specific Th2 clones showed identical results,[14] confirming the validity of the model using polyclonal populations pulsed with mitogens or mitogenic mAb.

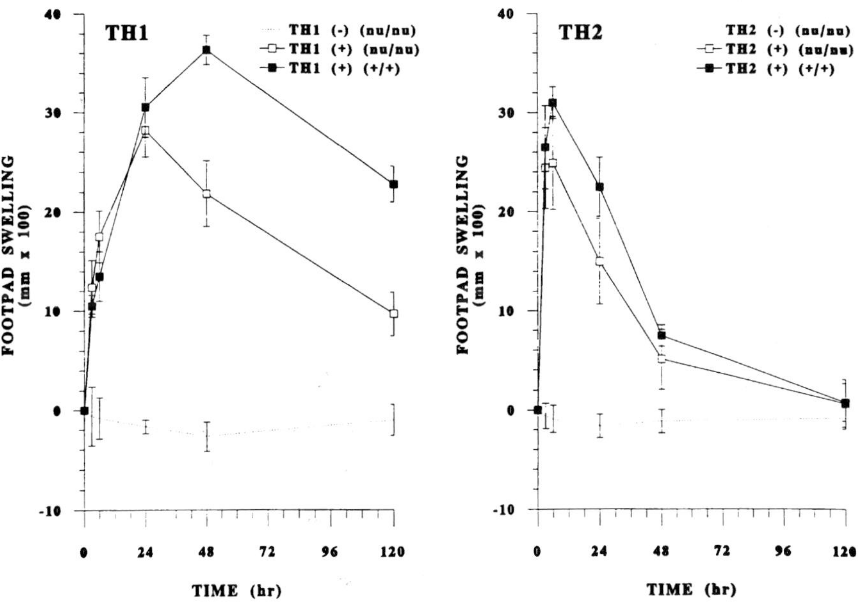

FIGURE 2 Th1- and Th2-mediated footpad swelling reactions in BALB/c *nu/nu* mice. Polyclonal Th1 and Th2 cells were pulsed with mAb 145-2C11 and injected into the hind paws of BALB/c (closed symbols) or BALB/c *nu/nu* (open symbols) mice (2.5×10^5/animal). Cells not previously pulsed with mAb 145-2C11 (no symbols) were injected as negative controls. Swelling was monitored over time, as indicated. Data are shown as mean ±SD, *n* = 5.

The time course of DTH2 responses was strikingly different from that of DTH reactions induced by Th1 cells, since strong swelling occurred very rapidly (within 2 to 3 h) following the injection of Th2 cells, peaked at 6 to 9 h, and then progressively decreased to reach baseline by 48 h. DTH2 was strongly dependent upon IL-4, since blocking concentrations of anti-IL-4 mAb or soluble IL-4 receptor (sIL-4R) given by ip injection also completely blocked tissue swelling in BALB/c mice; saturation of the sIL-4R with recombinant IL-4 prior to injection abrogated its inhibitory effect. IL-4 was required very early in the onset of the DTH2 swelling response, since ip administration of sIL-4R 2 h after the injection of Th2 cells had no inhibitory effect.[14] This is consistent with the very rapid induction of IL-4 release from polyclonal Th2 cells as well as from the alloantigen-specific (C3H anti-BALB/c) Th2 clone 2F11: increased IL-4 bioactivities could be measured in the supernatants as soon as 1 to 2 h after *in vitro* restimulation (Figure 3). The difference in the cytokine dependence of reactions elicited by Th1 and Th2 cells in BALB/c mice was quite clear-cut: neither anti-TNF nor mAb to IFN-γ affected DTH2 reactions, while DTH1 responses remained unaffected by sIL-4R or mAb to IL-4.

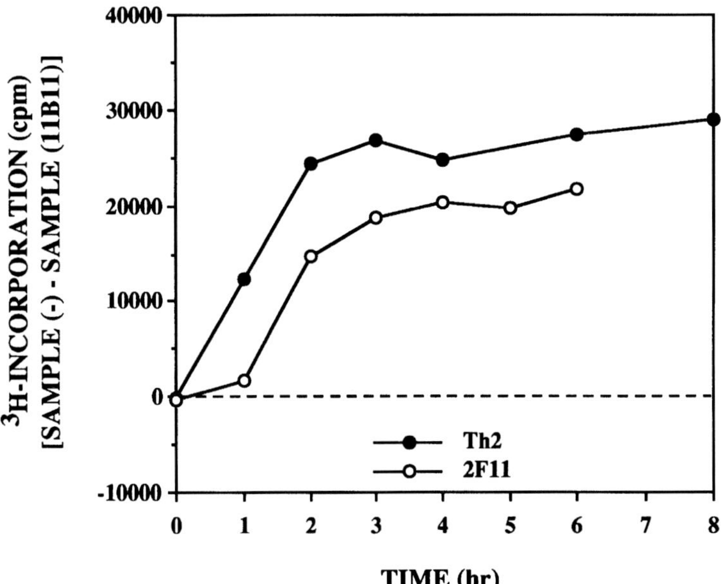

FIGURE 3   Kinetics of *in vitro* production of IL-4 by polyclonal Th2 cells (closed circles) and by the alloantigen-specific Th2 clone 2F11 (open circles). Polyclonal Th2 cells (5×10^5/ml) from BALB/c mice were restimulated with 1×10^6/ml syngeneic irradiated spleen cells and 1 µg/ml Con A. 2F11 cells were restimulated in a similar fashion with allogeneic BALB/c spleen cells in the absence of exogenous antigen. Supernatants were collected at the indicated times and tested for their capacity to sustain the IL-4-dependent growth of CT.4S cells. Data are shown as cpm (without inhibitor) – cpm (in the presence of saturating concentrations of anti-IL-4 mAb 11B11). Only mean data are shown; the SD were consistently <10% of mean values ($n = 3$).

Together, these results strongly suggested that Th1 and Th2 mediate tissue inflammation specifically via their prototypic cytokines. Interestingly, although distinct cytokines were responsible for DTH1 and DTH2 reactions, and although the swelling kinetics of the latter differed, the histological appearance of the lesions could not be distinguished: at 6 h after injection of either Th cell type, a polymorphonuclear leukocyte infiltrate was observed; at 48 h, mononuclear cells prevailed in both cases. It can therefore be speculated that different types of infiltrating cells play a major role in these two mechanisms of cutaneous inflammation; mononuclear cells such as macrophages might be activated by Th1 cells, whereas the earlier polymorphonuclear infiltrate might react to signals derived from Th2 cells. In this context, however, the role of IL-5, a prototypic Th2 cytokine, is unclear since we have not been able to demonstrate any eosinophil infiltration into lesions caused by the local injection of Th2 cells, and because ip injection of 1.5 mg of blocking anti-IL-5 mAb TRFK-5 had no effect on Th2-mediated swelling.

When C57BL/6 (H-2^b) strain mice were used instead of BALB/c (H-2^d) mice for DTH2 experiments, the injection of soluble IL-4 receptor (even at tenfold higher doses than required in BALB/c mice) was not sufficient to totally suppress the DTH2 reaction, although the amounts of IL-4 bioactivity released from Th2 cells of both strains were similar on a per cell basis. Instead, in C57BL/6 mice, anti-TNF antibody was required in addition to sIL-4R to completely suppress DTH2 reactions. Interstrain cellular transfers demonstrated that the TNF dependence was dictated by the donor T cells, but not by the recipient: BALB/c Th2 cells mediated strictly IL-4-dependent and C57BL/6 Th2 cells mediated both IL-4- and TNF-dependent inflammation, regardless of the recipient mouse strain. In experiments with intra-H-2 recombinant mice, the TNF dependence of DTH2 reactions correlated with a gene polymorphism segregating with the H-2D region of the MHC, which is in close vicinity of the gene encoding TNF-α. This TNF dependence correlated with the amount of TNF produced: Th2 cells from H-2D^b mice produced 5- to 30-fold higher levels of bioactive TNF than Th2 cells from H-2D^d mice.[15] Such a quantitatively differential TNF production (as found in T cells) has not been observed with macrophages.[43] These experiments suggested that a genomic polymorphism directing TNF expression may determine the participation of this cytokine in protective but also pathogenic immune reactions involving Th2 cells. Moreover, this may not only apply to Th2 cells, because Th1 cells from H-2^b haplotype mice also produced about tenfold more TNF than their counterparts from H-2^d mice, although the production in Th1 cells was generally much higher than in the corresponding Th2 cells.[15] We therefore think that the relatively high amounts of TNF produced by Th1 cells from both "high"- and "low"-TNF producer strains account for the involvement of this cytokine in DTH1 reactions in both types of strains; in contrast, the differential involvement of TNF in DTH2 may be explained as follows: (1) Th2 cells produced little, if any, TNF-β mRNA (K.M. Müller, Jean-François Arrighi, and C. Hauser, unpublished), and (2) regardless of the strain considered, Th1 cells always produced tenfold more bioactive TNF than Th2 cells.[15]

Although IL-4 has been ascribed to many down-regulatory immune functions, there are several lines of evidence demonstrating proinflammatory activities of IL-4. Transgeneic mice overexpressing IL-4 developed allergic-like blepharitis with a mononuclear cell infiltrate.[44] It has been described that tumor cells transfected with an expression vector for IL-4 induce an infiltrate composed of eosinophils.[45] As mentioned before, we have not observed eosinophils infiltrating the lesions induced by anti-CD3ε-pulsed Th cells. As anti-CD3ε is probably rapidly down-modulated, T cell cytokine production may last for only a short period in our system. Thus, short pulses of Th2 cytokine release may not last long enough to mobilize eosinophils into the tissue lesion. However, long-lasting IL-4 release *in vivo*, as observed with transfected tumor cells, may be capable of inducing blood eosinophils and tissue infiltration with eosinophils.[45] Accordingly, allergen-induced peribronchial inflammation and eosinophil as well as lymphocyte numbers in bronchoalveolar-lavage fluid were decreased in IL-4 deficient mice as compared to normal and mast cell-deficient controls.[46] Further evidence for a proinflammatory role of Th2 cells *in vivo* was obtained in a murine respiratory syncytial virus model: pulmonary lesions with infiltrating polymorphonuclear leukocytes could be prevented by a combined administration of mAb to IL-4 and IL-10.[47] These findings strengthen the concept that Th2 cells play an important role not only in humoral immunity, but also in T cell-dependent inflammatory reactions. It is well documented that allergic disorders such as atopic dermatitis and allergic asthma are associated with increased IgE levels; since IL-4 is an absolute requirement for IgE production, this correlation represents a sign of Th2 cell activation *in vivo* in these diseases. The expression of Th2-like cytokine patterns in skin from patients with a late phase skin reaction, in bronchoalveolar lavage from asthmatic, and in biopsies of nasal mucosa from patients with allergic rhinitis, as well as the cloning of cells with a Th2-like cytokine profile from atopics, have been reported.[17] Taken together, all these findings indicate that the activation of Th2-like cells is crucially important in atopic disorders.

IV. SUMMARY

Th cells are thought to be responsible for delayed-type hypersensitivity and possibly other inflammatory reactions. Through the release of distinct sets of cytokines, the Th cell subsets Th1 and Th2 exert characteristic and often mutually exclusive or antagonistic immune effector functions. We have documented and discussed recent findings on their respective proinflammatory functions *in vivo*. Th1 cells are often considered to mount inflammatory cellular immune responses, whereas Th2 cytokine profiles are generally thought to induce humoral immune responses and have been associated with inefficient elimination or infectious agents. We have recently developed an *in vitro* culture system that allows the short-term generation of polyclonal murine Th1 and Th2 cell populations. To assess the proinflammatory function of such polyclonal Th1 and Th2 cells *in vivo*, we tested them in a local adoptive transfer assay by subcutaneous injection into the skin of naive recipients. As expected, Th1 cells mediated local tissue inflammation that was entirely dependent on the two cytokines TNF and IFN-γ; however, in contrast to previous results obtained with selected clones, we found that Th2 cells were also capable of mediating local inflammatory reactions that depended on their prototypic cytokine IL-4, and, in high TNF-producing mouse strains, upon TNF-α. The swelling reactions mediated by Th2 cells peaked at 6 h and thus had kinetics strikingly different from those mediated by Th1 cells (maximum at 24 to 48 h); however, the histological pattern of infiltrating leukocytes induced by the two subsets was not distinguishable. It showed a polymorphonuclear infiltrate at 6 h and a predominance of mononuclear cells at 24 to 48 h. The cytokine dependence of Th1- or Th2-mediated inflammation was shown to be essentially dictated by the T helper cell donor. In contrast to the widely accepted belief that only Th1 cells can mediate delayed hypersensitivity reactions, our results demonstrate that other T cell-derived cytokine profiles may induce tissue inflammation with leukocytic infiltrates.

ACKNOWLEDGMENTS

We would like to thank Prof. J.-H. Saurat and Drs. S. Lisby, M. Lacour, I. Masouyé, G.E. Grau, P.-F. Piguet, and F. Aebischer, as well as F. Jaunin, J.-F. Arrighi, and M.-J. Cartier-Deldon for their contributions. This work was supported by grants from the Swiss National Foundation for Scientific Research (3.115-0.88, 32-27159.89, and 31-30930.91 to C.H.).

References

1. **Mosmann T.R., Cherwinski, H., Bond, M.W., Giedlin, M.A., and Coffman, R.L.,** Two types of murine helper T cell clone. I. Definition according to profiles of lymphokine activities and secreted proteins, *J. Immunol.*, 136, 2348, 1986.
2. **Cherwinski, H.M., Schumacher, J.H., Brown, K.D., and Mosmann, T.R.,** Two types of mouse helper T cell clone. III. Further differences in lymphokine synthesis between Th1 and Th2 clones revealed by RNA hybridization, functionally monospecific bioassays, and monoclonal antibodies, *J. Exp. Med.*, 166, 1229, 1987.

3. **Mosmann, T.R. and Coffman, R.L.,** Th1 and Th2 cells: different patterns of lymphokine secretion lead to different functional properties, *Annu. Rev. Immunol.*, 7, 145, 1989.

4. **Mosmann, T.R. and Moore, K.W.,** The role of IL-10 in crossregulation of Th1 and Th2 responses, *Immunol. Today*, 12, A49, 1991.

5. **Firestein, G.S., Roeder, W.D., Laxer, J.A., Townsend, K.S., Weaver, C. T., Hom, J.T., Linton, J., Torbett, B.E., and Glasebrook, A.L.,** A new murine CD4$^+$ T cell subset with an unrestricted cytokine profile, *J. Immunol.*, 143, 518, 1989.

6. **Röcken, M., Saurat, J.H., and Hauser, C.,** A common precursor for CD4$^+$ T cells producing IL-2 or IL-4, *J. Immunol.*, 148, 1031, 1992.

7. **Seder, R.A. and Paul, W.E.,** Acquisition of lymphokine-producing phenotype by CD4$^+$ T cells, *Annu. Rev. Immunol.*, 12, 635, 1994.

8. **Trinchieri, G.,** Interleukin-12 and its role in the generation of TH1 cells, *Immunol. Today*, 14, 335, 1993.

9. **Röcken, M., Müller, K.M., and Hauser, C.,** Induction of T helper cell subsets, *Springer Semin. Immunopathol.*, 13, 289, 1992.

10. **Röcken, M., Müller, K.M., Saurat, J.H., and Hauser, C.,** Lectin-mediated induction of IL-4 producing CD4$^+$ cell, *J. Immunol.*, 146, 577, 1991.

11. **Röcken, M., Müller, K.M., Saurat, J.H., Müller, I., Louis, J.A., Cerottini, J. C., and Hauser, C.,** Central role for TCR/CD3 ligation in the differentiation of CD4$^+$ T cells toward a Th1 or Th2 functional phenotype, *J. Immunol.*, 148, 47, 1992.

12. **Leo, O., Foo, M., Sachs, D.H., Samelson, L.E., and Bluestone, J.A.,** Identification of a monoclonal antibody specific for a murine T3 polypeptide, *Proc. Natl. Acad. Sci. U.S.A.*, 84, 1374, 1987.

13. **Donovan, J. and Brown, P.,** Parenteral injections: footpad injection of the mouse, in *Current Protocols in Immunology*, Coligan, J.E., Kruisbeek, A.M., Margulies, D.H., Shevach, E.M., and Strober, W., Eds., Greene Publishing and Wiley-Interscience, New York, 1991, 1.6.8.

14. **Müller, K.M., Jaunin, F., Masouyé, I., Saurat, J.H., and Hauser, C.,** Th2 cells mediate IL-4-dependent local tissue inflammation, *J. Immunol.*, 150, 5576, 1993.

15. **Müller, K.M., Lisby S., Arrighi, J.F., Grau, G.E., Saurat, J.H., and Hauser, C.,** H-2D haplotype-linked expression and involvement of TNF-α in Th2 cell-mediated tissue inflammation, *J. Immunol.*, 153, 316, 1994.

16. **Hu-Li, J., Ohara, J., Watson, C., Tsang, W., and Paul, W.E.,** Derivation of a T cell line that is highly responsive to IL-4 and IL-4 (CT.4R) and of an IL-2 hyporesponsive mutant of that line (CT.4S), *J. Immunol.*, 142, 800, 1989.

17. **Müller, K.M. and Hauser, C.,** The biological and clinical significance of helper T-cell subsets in the human, *Exp. Dermatol.*, 1, 161, 1992.

18. **Ju, S.T., Ruddle, N.H., Strack, P., Dorf, M.E., and DeKruyff, R.H.,** Expression of two distinct cytolytic mechanisms among murine CD4 subsets. *J. Immunol.*, 144, 23, 1990.

19. **Stalder, T., Hahn, S., and Erb, P.,** Fas antigen is the major target molecule for CD4$^+$ T cell-mediated cytotoxicity, *J. Immunol.*, 152, 1127, 1994.

20. **Del Prete, G., De Carli, M., Almerigogna, F., Giudizi, M.G., Biagiotti, R., and Romagnani, S.,** Human IL-10 is produced by both type 1 helper (Th1) and type 2 helper (Th2) T cell clones and inhibits their antigen-specific proliferation and cytokine production, *J. Immunol.*, 150, 353, 1993.

21. **Stevens, T.L., Bossie, A., Sanders, V.M., Fernandez-Botran, R., Coffman, R. L., Mosmann, T.R., and Vitetta, E.S.,** Regulation of antibody isotype secretion by subsets of antigen-specific helper T cells, *Nature (London)*, 334, 255, 1988.

22. **Hauser, C., Snapper, C.M., Ohara, J., Paul, W.E., and Katz, S.I.,** T helper cells grown with hapten-modified cultured Langerhans cells produce IL-4 and stimulate IgE synthesis, *Eur. J. Immunol.*, 19, 245, 1989.

23. **Lehmann, P.V., Schumm, G., Moon, D., Hurtenbach, U., Falcioni, F., Muller, S., and Nagy, Z.A.,** Acute lethal graft-versus-host reaction induced by major histocompatibility complex class II-reactive T helper cell clones, *J. Exp. Med.*, 171, 1485, 1990.

24. **Tary-Lehmann, M., Rolink, A.G., Lehmann, P.V., Nagy, Z.A., and Hurtenbach, U.,** Induction of graft versus host-associated immunodeficiency by CD4$^+$ T cell clones, *J. Immunol.* 145, 2092, 1990.

25. **Fowler, D.H., Kurasawa, K., Husebekk, A., Cohen, P.A., and Gress, R.E.,** Cells of Th2 cytokine phenotype prevent LPS-induced lethality during murine graft-versus-host reaction. Regulation of cytokines and CD8$^+$ lymphoid engraftment, *J. Immunol.*, 152, 1004, 1994.

26. **Fanslow, W.C., Clifford, K.N., Park, L.S., Rubin, A.S., Voice, R.F., Beckmann, M.P., and Widmer, M.B.,** Regulation of alloreactivity *in vivo* by IL-4 and the soluble IL-4 receptor, *J. Immunol.*, 147, 535, 1991.

27. **Doutrelepont, J.M., Moser, M., Leo, O., Abramowicz, D., Vanderhaegen, M.L., Urbain, J., and Goldman, M.,** Hyper IgE in stimulatory graft-*versus*-host disease: role of interleukin-4, *Clin. Exp. Immunol.*, 83, 133, 1991.

28. **Goldman, M., Druet, P., and Gleichmann, E.,** Th2 cells in systemic autoimmunity: insights from allogeneic diseases and chemically-induced autoimmunity, *Immunol. Today*, 12, 223, 1991.

29. **Ochel, M., Vohr, H.W., Pfeiffer, C., and Gleichmann, E.,** IL-4 is required for the IgE and IgG1 increase and IgG1 autoantibody formation in mice treated with mercuric chloride, *J. Immunol.*, 146, 3006, 1991.

30. **Snapper, C.M., and Paul, W.E.,** Interferon-γ and B cell stimulatory factor-1 reciprocally regulate Ig isotype production, *Science*, 236, 944, 1987.

31. **Urban, J.F., Jr., Madden, K.B., Svetic, A., Cheever, A., Trotta, P.P., Gause, W.C., Katona, I.M., and Finkelman, F.D.,** The importance of Th2 cytokines in protective immunity to nematodes, *Immunol. Rev.*, 127, 205, 1992.

32. **Wang, Z.E., Reiner, S.L., Zheng, S., Dalton, D.K., and Locksley, R.M.,** CD4$^+$ effector cells default to the Th2 pathway in interferon γ-deficient mice infected with *Leishmania major, J. Exp. Med.*, 179, 1367, 1994.

33. **Afonso, L.C.C., Scharton, T.M., Vieira, L.Q., Wysocka, M., Trinchieri, G., and Scott, P.,** The adjuvant effect of interleukin-12 in a vaccine against *Leishmania major, Science*, 263, 235, 1994.

34. **Scott, P. and Kaufmann, S.H.E.,** The role of T-cell subsets and cytokines in the regulation of infection, *Immunol. Today*, 12, 346, 1991.

35. **Salgame, P., Abrams, J.S., Claybeger, C., Goldstein, H., Convit, J., Modlin, R.L., and Bloom, B.R.,** Differing lymphokine profiles of functional subsets of human CD4 and CD8 T cell clones, *Science*, 254, 279, 1991.

36. **Yamamura, M., Uyemura, K., Deans, R.J., Weinberg, K., Rea, T.H., Bloom, B.R., and Modlin, R.L.,** Defining protective responses to pathogens: cytokine profiles in leprosy lesions, *Science*, 254, 277, 1991.

37. **Clerici, M. and Shearer, G.M.,** A Th1→Th2 switch is a critical step in the etiology of HIV infection, *Immunol. Today*, 14, 107, 1993.

38. **Cher, D.J. and Mosmann, T.R.,** Two types of murine helper T cell clone. II. Delayed-type hypersensitivity is mediated by TH1 clones, *J. Immunol.*, 138, 3688, 1987.

39. **Hauser, C.,** Cultured epidermal Langerhans' cells activate T effector cells for contact sensitivity, *J. Invest. Dermatol.*, 95, 436, 1990.

40. **Fong, T.A.T. and Mosmann, T.R.,** The role of IFN-γ in delayed-type hypersensitivity mediated by Th1 clones, *J. Immunol.*, 143, 2887, 1989.

41. **Wong, R.L., Ruddle, N.H., Padula, S.J., Lingenheld, E.G., Bergman, C. M., Rugen, R.V., Epstein, D.I., and Clark, R.B.,** Subtypes of helper cells. Non-inflammatory type 1 helper T cells, *J. Immunol.*, 141, 3329, 1988.

41a. **Müller, K.M., Röcken, M., Carlberg, C., and Hauser, C.,** *J. Invest. Dermatol.*, in press.

42. **Tsicopoulos, A., Hamid, Q., Varney, V., Ying, S., Moqbel, R., Durham, S. R., and Kay, A.B.,** Preferential messenger RNA expression of Th1-type cells (IFN-γ⁺, IL-2⁺) in classical delayed-type (tuberculin) hypersensitivity reactions in human skin, *J. Immunol.*, 148, 2058, 1992.

43. **de Kossodo, S. and Grau, G.E.,** Profiles of cytokine production in relation with susceptibility to cerebral malaria, *J. Immunol.*, 151, 4811, 1993.

44. **Tepper, R.I., Levinson, D.A., Stanger, B.Z., Campos-Torres, J., Abbas, A. K., and Leder, P.,** IL-4 induces allergic-like inflammatory disease and alters T cell development in transgenic mice, *Cell*, 62, 457, 1990.

45. **Tepper, R.I., Coffman, R.L., and Leder, P.,** An eosinophil-dependent mechanism for the antitumor effect of interleukin-4, *Science*, 257, 548, 1992.

46. **Brusselle, G.G., Kips, J.C., Tavernier, J.H., van der Heyden, J.G., Cuvelier, C.A., Pauwels, R.A., and Bluethmann, H.,** Attenuation of allergic airway inflammation in IL-4 deficient mice, *Clin. Exp. Allergy*, 24, 73, 1994.

47. **Connors, M., Giese, N.A., Kulkarni, A.B., Firestone, C.Y., Morse, H. C., III, and Murphy, B.R.,** Enhanced pulmonary histopathology induced by respiratory syncytial virus (RSV) challenge of formalin-inactivated RSV-immunized BALB/c mice is abrogated by depletion of interleukin-4 (IL-4) and IL-10, *J. Virol.*, 68, 5321, 1994.

4 The Use of Scalp Grafts onto Nude Mice as a Model for Human Hair Growth: Is There Something New for Hair Growth Drug Screening Programs?

Dominique Van Neste

TABLE OF CONTENTS

I.   INTRODUCTION

During recent years, interest in human scalp hair biology has been stimulated by at least two factors: (1) substantial improvement of the technology in cutaneous biology at the cellular and tissular levels and (2) significant input of "outsiders," i.e., scientists not primarily involved in skin research but attracted to the field from other areas of biological sciences.[1]

Concurrently, the interest of clinicians was triggered by the discovery of compounds evaluated in subjects with male pattern baldness that showed some efficacy in reverting a variable proportion of vellus follicles into terminal hair follicles.[3,14]

There is a need for clinically relevant experimental models. Indeed, it is not yet firmly established whether the information that is being generated by *in vitro* experiments or by using animal models is always clinically relevant and correlates with results obtained in clinical trials with human hair or scalp diseases. Therefore there is much room left for outlining the clinical relevance of existing laboratory tests and/or for developing a model of normal and diseased human scalp that would be ready for testing compounds in the laboratory, i.e., without the ethical aspects as soon as human subjects become involved. Indeed, all test compounds will need, at one stage or another, to be tested on human scalp *in vivo*. Moreover, many pharmaceutical and/or cosmetic laboratories would appreciate having a chance to determine at very early stages—ideally before launching expensive drug toxicity and clinical development program—whether newly synthesized compounds do influence human hair growth.

Today, it is mandatory to increase the knowledge of factors able to modulate hair growth when the human hair follicle is maintained in laboratory conditions within or outside its natural environment (including sebaceous glands, sweat glands, variable and constant portion of the hair follicle, follicular opening, and interfollicular epidermis).

Therefore knowledge of the primary requirements in terms of clinical efficacy is essential. The dynamics of hair productivity (or hair output) resulting clinically into the "hairiness" of any given skin area is determined by four variables:

- the absolute number of hair follicles that are functionally active,
- the duration of the anagen phase,
- the linear hair growth rate, and
- the thickness of the hair fibers.

These four "analytical" factors can be measured individually and mathematically grouped to generate a more "global" compound index of hair growth in health and disease (Van Neste and Rushton, unpublished data).

Human scalp disorders and alopecia are due to changes in one, or a combination of several hair growth variables as defined above. As a result, a model that would reflect each of these variables would be considered an ideal candidate for drug screening programs.

II. HUMAN HAIR FOLLICLE GRAFTS ONTO NUDE MICE: AN IDEAL MODEL FOR STUDYING HUMAN SCALP HAIR DISORDERS?

Autologous and heterologous grafts have been made using the "nude" mouse as a recipient. Data about human and other species' skin grafted onto "nude" mice can be found in earlier review papers.[5,17] The main points of relevance to the hair follicle biology are the following:

- Human skin transplants keep their original structural and functional integrity.
- Structurally, the hairs look very much like human hair and they clearly differ from the rare, minute, thin, short, and unpigmented mouse hairs.
- When disease is due to intrinsic modifications of the skin, the pattern persists long after grafting.
- When inflammatory skin diseases are evaluated, the infiltrate is diluted after a time and the skin samples "normalize" subsequently.

Before describing this model in relation to hair growth, it is necessary to examine the technical difficulties, i.e., rate of success of transplantation and survival of the samples. In subsequent sections we will discuss other aspects, such as the use of embryonic scalp specimens, and address questions related to the subsequent continuous synthesis of hair, its shape, its protein content (in normal and genetically deficient human scalp samples), and others. To put all the information contained in this review into the appropriate perspective, it is necessary to know that normal hair follicles maintained *in vitro* start to die off as soon as the fifth day in culture.[7,11]

A. RATE OF SUCCESS OF TRANSPLANTATION OF HUMAN SCALP ONTO "NUDE" MICE

The rate of graft take clearly is a matter of experience. Indeed, we improved our rate of graft take from 30 to almost 50% during the past 5 years. It is not established yet whether technical aspects (size of grafts, transport, nutrients, and medium temperature) play a significant role in the survival of grafts, but, during the first 2 months, there is almost 50% elimination of the grafts in our experience and that of others. Indeed, using 2-mm punch grafts Gilhar and Krueger[6] obtained an 88% rate of success of graft take, which dropped to 42% by day 78. Also, unless extreme care is taken, some hairs grow underneath nude skin (Figures 1 and 2). Another unsolved question is how many follicles are maintained in taken grafts. Finally, it is not known whether the cycling activity is maintained, and, if so, whether cycling is still under hormonal control. These questions are actually under consideration in our laboratory and will be mentioned later (see Section II.B).

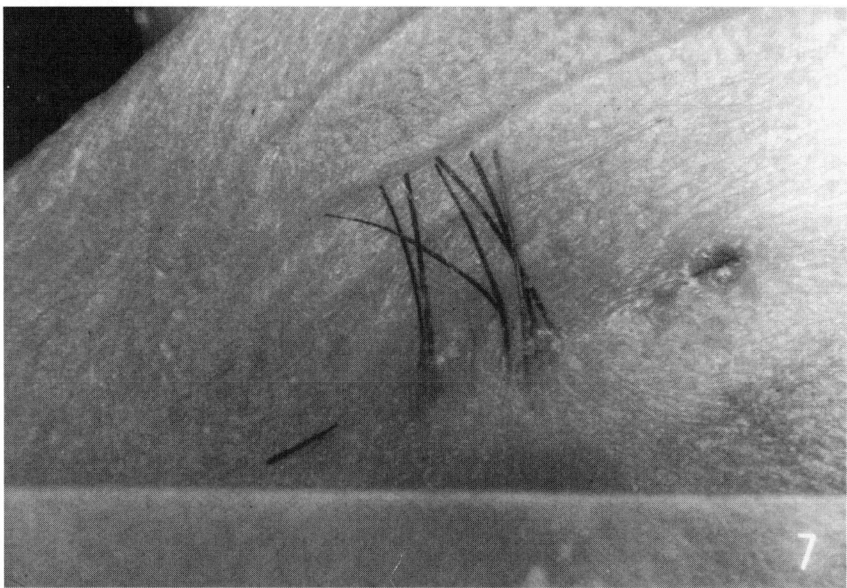

FIGURE 1 Normal human scalp hair follicle grafts showing human hairs growing at the surface of nude mouse skin.

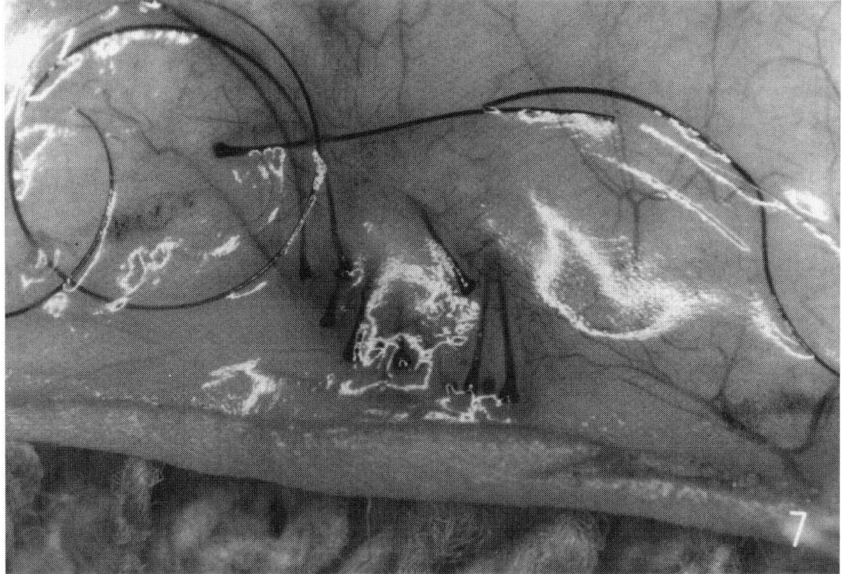

FIGURE 2 Flap showing the hair follicle roots of the same specimen as shown in Figure 1. Some hairs are growing into the dermal–subcutaneous junction.

B.   EMBRYOLOGY AND MORPHOGENESIS

Human fetal skin grafts onto the "nude" mice have been used for the study of follicle morphogenesis. Almost 80% of the grafted specimens took and gave adequate hair follicle development.[8] The observation of human fetal scalp grafts demonstrated that the development of epidermis, dermis, eccrine glands, and pilosebaceous apparatus continued in an orderly and timely fashion when transplanted onto "nude" mice. Estimated gestation age (EGA) did not play a significant role in appendageal development. This contrasted sharply with the results of *in vitro* maintenance of human fetal scalp. Indeed, no follicular organogenesis existed if samples were taken before the 35th day of EGA. At 65 to 75 days EGA, full follicular development took place because, at that time, induction had already occurred *in vivo*, and the process of morphogenesis simply needed to proceed *in vitro*. Most of these grafts have been placed subcutaneously or beneath the kidney capsule.[8]

When the subcutaneously implanted scalp grafts were exposed to the surface,[12] the majority of the specimens demonstrated a more advanced development over age-matched controls *in utero* or control grafts remaining in a subcutaneous situation. The consequences of air exposure in this model are very similar to the acceleration of epidermal development in preterm infancy and the accelerated terminal type differentiation of epidermis when epidermal cell cultures are exposed to air *in vitro*.

When fetal scalp samples were transplanted, we and others (Zareba et al., personal communication, 1992) noticed that the density of hair follicles (based on hairs produced per centimeter square) was very high (Figures 3 and 4), and certainly much higher than in the case of fully differentiated scalp of young or old subjects.[4] This might be ascribed to the fact that scalp expansion had not yet occurred at that stage of the fetal development. The next question that could be addressed is related to the gene expression and, more specifically, to the continuous protein synthesis process of hair compounds in this model.

C.   ANALYSIS OF PROTEIN COMPOSITION IN NORMAL SCALP AND GENETICALLY DEFICIENT HAIR FIBER PRODUCTION: COMPARATIVE STUDIES WITH TRICHOTHIODYSTROPHY (TTD AND TTD VARIANTS)

The regulation of gene expression in the hair shafts produced in this model has only recently been addressed. A noninvasive way to explore the gene expression is to analyze protein composition in the hairs merging at the surface of the human scalp grafts. Amino acid composition is a reliable way to do so on a small amount of material. Among the many different hair dysplasias, we had the opportunity to evaluate the amino acid composition of hairs taken from

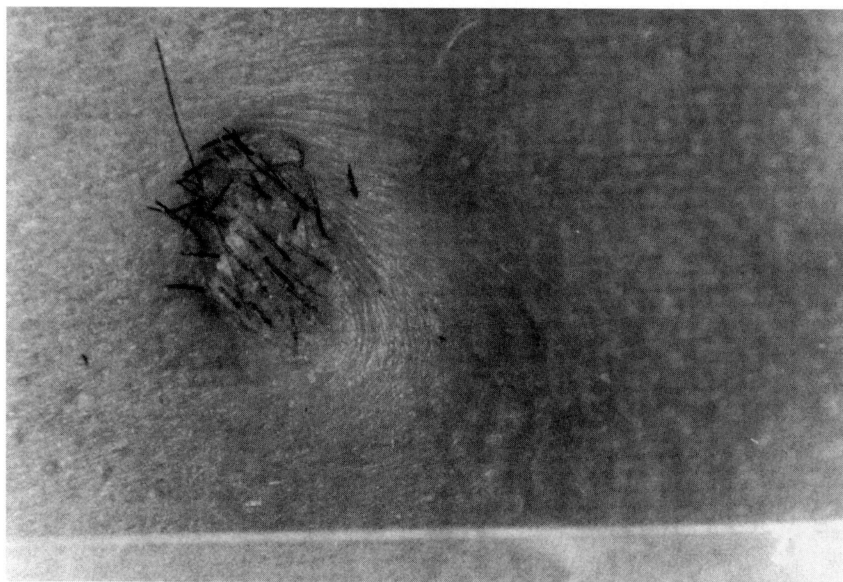

FIGURE 3   Phototrichogram study of trichothiodystrophic hairs growing at the surface of a grafted fetal scalp specimen. This view shows the hairs immediately after clipping. (From de Brouwer et al.,[4] with permission.)

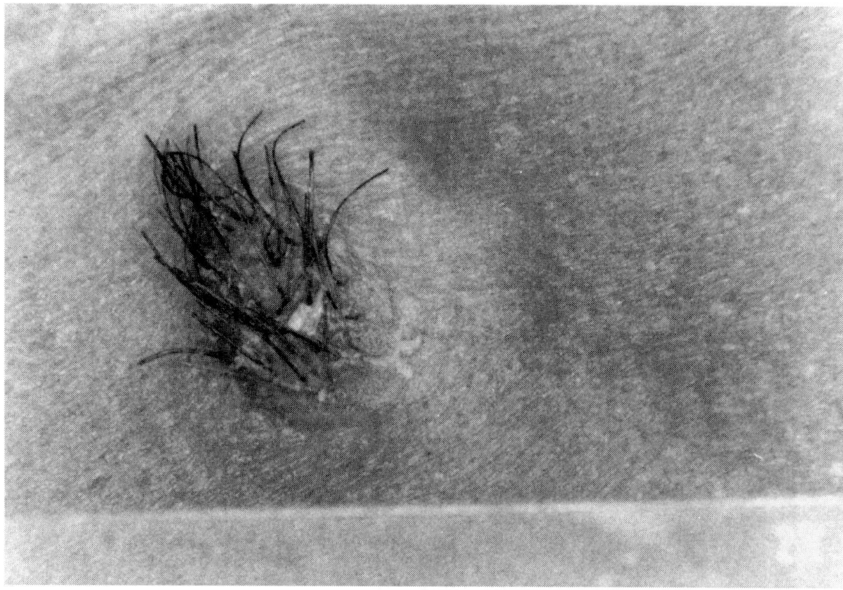

FIGURE 4   Phototrichogram study of trichothiodystrophic hairs growing at the surface of a grafted fetal scalp specimen. This view shows the same hairs as in Figure 3 but 4 days after clipping. (From de Brouwer et al.[4] with permission.)

healthy controls and patients with trichothiodystrophy (TTD) and related deviations. We also had the opportunity to obtain TTD scalp samples and to monitor the production of hair fibers on grafted specimens.[4,19]

Analyzing the amino acid composition of the hairs produced by scalp grafts onto "nude" mice and comparing the results with those obtained on hairs collected *in situ*, it appeared that the protein expression was similar after grafting in both normal and TTD scalp grafts. As TTD and TTD variants are due to abnormal expression of the genes encoding for high-sulfur proteins, it is clear that transplanting of scalp grafts onto "nude" mice can be an appropriate laboratory model for testing genetic hair diseases.

Furthermore, we hypothesized that the model could be effective for running assays with *in vivo* gene transfer to the human skin as has been demonstrated with epidermal cell culture transplants for the control of expression after *in vitro* gene transfer experiments.[10,16] This is especially interesting from an ethical point of view.

D. GRAFTING OF INFLAMED SCALP: THE EXAMPLE OF ALOPECIA AREATA

Alopecia areata appears to be initiated by an inflammatory insult that precipitates anagen follicles prematurely into catagen and telogen. Then, when the bald patches become established, immature anagen follicles not developing beyond stage IV remain present. At this stage, there is no visible hair at the scalp surface. The inflammatory cell infiltrate is mainly composed of activated T lymphocytes associated with macrophages and Langerhans cells.[9] This process results eventually into miniaturized follicles in long standing lesions.

When scalp skin grafts are obtained from patients with alopecia areata and alopecia universalis, and transplanted onto "nude" mice, there is hair regrowth.[5] This demonstrates that the process is secondary to a factor that is systemic and host specific. Injecting the patient's serum into the grafted mice did not affect the hair growth process. This implies that serum factors are probably secondary manifestations of the alopecia rather than primary factors having a significant role in modulating the hair follicle growth in alopecia areata. Conversely, as indicated by immunopathology, it is most probable that cellular immune response and especially T cell subsets are involved and play a primary role in the pathogenesis of alopecia areata.

A series of drug trials using alopecia areata scalp grafts onto "nude" mice have been reviewed. It has been reported that both oral and topical application of cyclosporin A would cause hair regrowth in grafted alopecia areata scalp samples. However, the specificity of the result is questionable. Indeed, there is a spontaneous dilution of the inflammatory infiltrate form the alopecia areata scalp samples when they are implanted into the "nude" mouse. The fact that cyclosporin A induces acceleration of hair growth *in vitro* and in normal as well as in "nude" mice is further evidence that the effect of the drug may not be stimulating hair growth via clearance of the inflammatory infiltrate in alopecia areata grafts.[2,15] Hence models that would be indicative of therapeutic properties of compounds worth being tried in inflammatory scalp disorders of human subjects are still lacking.

E. GRAFTING OF NONINFLAMED SCALP: THE EXAMPLE OF ANDROGEN-DEPENDENT ALOPECIA

In androgen-dependent alopecia, there is little evidence of a specific inflammatory process. Conversely, the gradual miniaturization of the terminal type into vellus type follicles is highly characteristic. The associated shortening of the hair cycle and the reduction in hair pigmentation result in less visible vellus hairs.[20] Androgen-dependent alopecia is known to affect specific areas of the human scalp, especially in males, but also in females though with a different pattern. This pattern as described in the original reports by Hamilton is expressed only when androgens are present. In the absence of androgens there is no initiation of—or if suppression occurred after puberty, there is no further progression of—the balding process inasmuch as the subjects are genetically predisposed.[3,14] The regional control has been elegantly highlighted by Nordström's experiments using autologous transplantation: when affected scalp sites are transplanted to other body sites there is progression of the balding pattern at the same pace in the donor area and in the transplant.[13]

On balding scalp samples that were transplanted onto "nude" mice, we demonstrated that hair growth continued and could be evaluated over a 6-month period. The distribution of diameters of the hairs collected from the grafted specimens was similar to that of hairs collected *in situ*, i.e., on the scalp of subjects affected with androgen-dependent alopecia. However, the linear hair growth rate (millimeters per day) was decreased in grafts.[18] The decrease equally affected terminal and vellus hairs. This indicates that the systemic environment, i.e., nutrients provided by the "nude" system, most probably is slightly deficient as compared with that provided in the humans. In our hands (unpublished data), balding scalp specimens usually gave rise to a single hair cycle of short duration (less than 3 months). Incipient recycling is clearly demonstrated histologically (Figure 5),[17] and this eventually results in new hairs visible at the scalp surface. We recently observed a short second cycle in 2 out of 10 specimens that could be observed at least during a 6-month period after grafting (unpublished data).

F. TOXICOLOGY AND PHARMACOLOGY OF TRANSPLANTED HUMAN HAIR FOLLICLES

There is preliminary evidence from the toxicological point of view that the grafted human hair follicles react in a situation very similar to the *in vivo* situation.[22] There are very few published studies demonstrating the usefulness of human scalp transplanted to "nude" mice for evaluating drug compounds

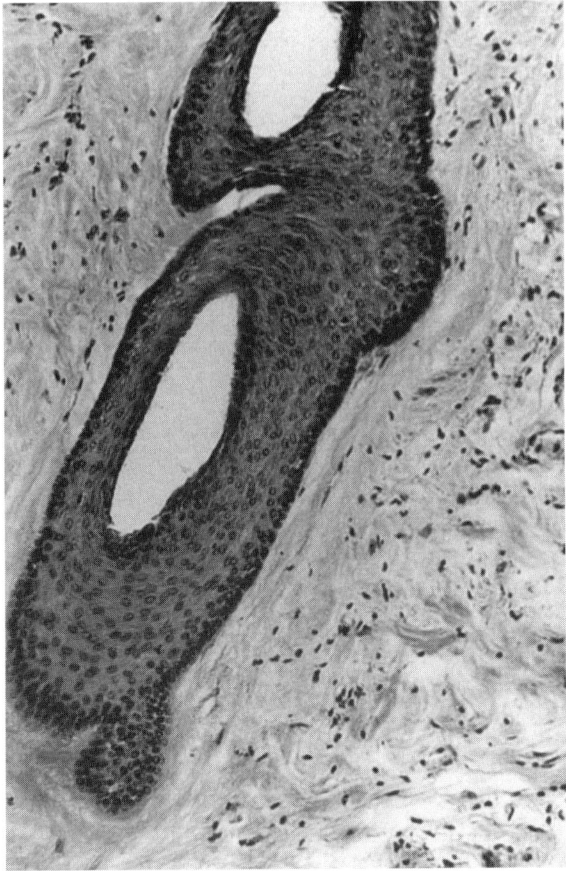

FIGURE 5   Hair cycle initiation in human hair follicle grafts. The epithelial budding is the earliest step in the newly developing hair follicle.

able to stimulate or to inhibit hair growth. At least in theory, this laboratory model appears as highly advantageous for drug discovery programs. Indeed, those scalp specimens affected with the disease to which the drug is targeted can be tested before expensive in depth toxicology programs have been launched. If a compound significantly enhances hair growth or hair regrowth on the model a decision can be made to go on with full drug development programs.

G.   THE MICROENVIRONMENT OF THE HAIR FOLLICLE

At present we know from transplantation of split skin grafts and total thickness scalp grafts that the sebaceous gland is present and remains active in a number of scalp samples (Figure 6; more references in Van Neste et al.[17]).

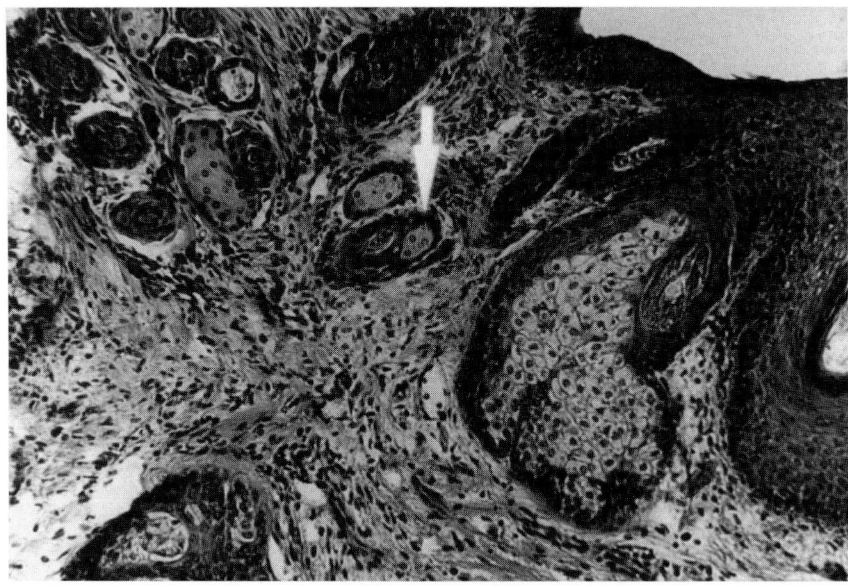

FIGURE 6   Sebaceous gland from mouse skin (left side, arrow) is miniature as compared with human sebaceous gland (right side).

Preliminary evaluation of the sebum output did not generate relevant data because of the presence of mouse sebaceous glands. A more specific model has to be developed for evaluation of human sebum production on the long term. Occasionally when scalp samples were taken next to cysts, such cysts also developed from the implanted follicles (Figures 7 and 8). Once again such observations illustrate that the pathological processes continued as they would do if the follicles were left in the scalp *in situ*.

III. FUTURE OF THE MODEL

There is preliminary evidence from various sources that human scalp graft onto the "nude" mouse is a reasonable experimental approach for the study of hair follicle physiology and pathology. It remains to be established whether in terms of industrial development it really answers all the questions that are clinically relevant for drug screening programs.

ACKNOWLEDGMENTS

My thanks to Bernadette de Brouwer who provided help during the preparation of figures and editing of the manuscript.

FIGURE 7   Clinical aspect of epithelial cyst developing from human scalp graft.

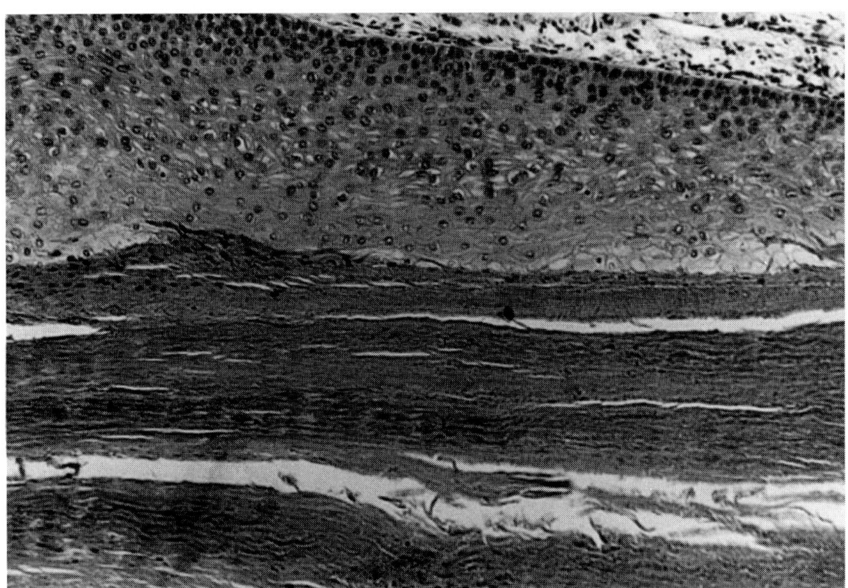

FIGURE 8   Histology of epithelial cyst showing thick multilayered horny layer inside the cyst (hemalum–eosin–saffron).

References

1. **Baden, H.P.,** Where we have been and where we should be. Clinical relevance of the molecular and structural biology of hair, in *The Molecular and Structural Biology of Hair,* Stenn, K.S., Messenger, A.G., and Baden, H.P., Eds, New York Academy of Sciences, New York, 1991, 435.

2. **Buhl, A.E., Waldon, D.J., Miller, B.F., and Brunden, M.N.,** Differences in activity of minoxidil and cyclosporin A on hair growth in nude and normal mice, *Lab Invest,* 62, 104, 1990.

3. **Dawber, R.P.R. and Van Neste, D.,** *Hair and Scalp Disorders,* Martin Dunitz, London, 1995.

4. **de Brouwer, B., Föhles, J., and Van Neste, D.,** Human hair production by scalp samples grafted onto nude mice. Biochemical data on normal human hair and the genetic defect trichothiodystrophy, *J. Dermatol. Sci.,* 7, s39, 1994.

5. **Gilhar, A. and Etzioni, A.,** The nude mouse model for the study of human disorders, *Dermatology,* 189, 5, 1994.

6. **Gilhar, A. and Krueger, G.G.,** Hair growth in scalp grafts from patients with alopecia areata and alopecia universalis grafted onto nude mice, *Arch. Dermatol.,* 123, 44, 1987.

7. **Harmon, C.S. and Nevins, T.D.,** Hair fibre production by human hair follicles in whole-organ culture, *Br. J. Dermatol.,* 130, 415, 1994.

8. **Holbrook, K.A. and Minami, S.I.,** Hair follicle embryogenesis in the human. Characterization of events in vivo and in vitro, in *The Molecular and Structural Biology of Hair,* Stenn, K.S., Messenger, A.G., and Baden, H.P. Eds., New York Academy of Sciences, New York, 1991, 167.

9. **Hordinsky, M.K.** Alopecia areata, in *Disorders of Hair Growth, Diagnosis and Treatment,* Olsen, E.A., Ed., McGraw-Hill, New York, 1993, 195.

10. **Jensen, U.B., Jensen, T.G., Jensen, P.K.A., Rygaard, J., Hansen, B.S., Fogh, J., Kolvraa, S., and Bolund, L.,** Gene transfer into cultured human epidermis and its transplantation onto immunodeficient mice. An experimental model for somatic gene therapy, *J. Invest. Dermatol.,* 103, 391, 1994.

11. **Jones, L.N., Fowler, K.J., Marshall, R.C., and Ackland, M.L.,** Studies of developing human hair shaft cells in vitro, *J. Invest. Dermatol.,* 90, 58, 1988.

12. **Lane, A.T., Scott, G.A., and Day, K.H.,** Development of human fetal skin transplanted to the nude mouse, *J. Invest. Dermatol.,* 93, 787, 1989.

13. **Nordström, R.E.A.,** Synchronous balding of scalp and hair-bearing grafts of scalp transplanted to the skin of the arm in male pattern baldness, *Acta Dermatol. Venereol. (Stockholm),* 59, 266, 1979.

14. **Olsen, E.A.,** Androgenetic alopecia, in *Disorders of Hair Growth, Diagnosis and Treatment,* Olsen, E.A., Ed., McGraw-Hill, New York, 1993, 257.

15. **Sawada, M., Terada, N., Taniguschi, H., Tateishi, R., and Mori, Y.,** Cyclosporin A stimulates hair growth in nude mice, *Lab Invest.,* 56, 684, 1987.

16. **Setoguchi, Y., Jaffe, H.A., Danel, C., and Crystal, R.G.,** Ex vivo and in vivo gene transfer to the skin using replication-deficient recombinant adenovirus vectors, *J. Invest. Dermatol.,* 102, 415, 1994.

17. **Van Neste, D., Warnier, G., Thulliez, M., and Van Hoof, F.,** Human hair follicle grafts onto nude mice: morphological study, in *Trends in Human Hair Growth and Alopecia Research*, Van Neste, D., Lachapelle, J.M., and Antoine, J. L., Eds, Kluwer Academic Publisher, Dordrecht, the Netherlands, 1989, 117.

18. **Van Neste, D., de Brouwer, B., and Dumortier, M.,** Reduced linear hair growth rates of vellus and of terminal hairs produced by human balding scalp grafted onto nude mice, in *The Molecular and Structural Biology of Hair*, Stenn, K.S., Messenger, A.G., and Baden, H.P., Eds., New York Academy of Science, New York, 1991, 480.

19. **Van Neste, D.J.J., Gillespie, J.M., Marshall, R.C., Taieb, A., and de Brouwer, B.,** Morphological and biochemical characteristics of trichothiodystrophy-variant are maintained after scalp grafts onto nude mice, *Br. J. Dermatol.*, 128, 384, 1993.

20. **Whiting, D.A.,** The value of horizontal sections of scalp biopsies, *J. Cutan. Aging Cosm. Dermatol.*, 1, 165, 1990.

21. **Zareba, G., Goldsmith, L.A., Pittman, K., and Clarkson, T.W.,** Characteristics of human hair grown on nude mice, European Hair Research Meeting, Berlin, October 2–3, 1992.

22. **Zareba, G., Goldsmith, L.A., and Clarkson, T.W.,** Development of human scalp/nude mice model to study biological monitoring of toxic substances undergoing accumulation into hair, European Hair Research Meeting, Stockholm, October 1–2, 1993.

5 Objective Measurements of Skin Color Using Reflectance Spectroscopy

Peter H. Anderson

TABLE OF CONTENTS

I. INTRODUCTION

The use of objective techniques was delayed in dermatology research compared to other fields of medicine and biophysics. This phenomenon can be explained by several factors. When immediately grading different color intensities the human eye is extremely sensitive and early attempts to measure skin colors objectively were disappointing. Second, visual color perception often intergrates the reaction size and the degree of edema, probably explaining why visual estimations often are equal to objective measuring techniques. Most bioengineering equipment measures a small area in the center of a response,[1-6] whereas visual impression integrates several conditions and the coherence by the trained eye and finger becomes a very powerful tool. However, in recent decades several noninvasive and objective measuring procedures have been introduced and increased our knowledge of several skin conditions.[1,2,7]

A noninvasive bioengineering system for objective cutaneous measurements must be nontraumatic and noninterfering with measured skin parameters. Furthermore, the device must be accurate, reliable, and show highly reproducible and informative measurements. Reflectance spectroscopic measurements were introduced into dermatology in 1939 by Edwards and Duntley,[8] but sparsely used in scientific investigations until the last two decades. After the introduction of fiber optics and faster scanners reflectance spectroscopy was considered a dependable and precise method for determination of cutaneous color variations. The technique was utilized to monitor several dermatological treatment effects and to study cutaneous biophysics.[1,9-11]

II. REFLECTANCE SPECTROSCOPIC EQUIPMENT

Reflectance spectroscopy of the skin can be performed in the wavelength range from ultraviolet to the infrared.[12-14] Most studies have been performed in the near ultraviolet and visible ranges. The skin is irradiated with light ranging from the upper part of the ultraviolet (350 nm) to the beginning of the infrared part (800 nm) of the spectrum (Figure 1).[1] The irradiation may either be polychromatic (white) with monochromatic detection or monochromatic irradiation with polychromatic light sampling. To reduce the influence of the ambient light most systems today uses polychromatic irradiation with monochromatic detection. The skin is illuminated through a flexible light guide (fiber) and the backscattered light reflected off the skin surface from the different epidermal and dermal layers is collected by a light integrating system. Through similar fibers the reflected light is guided back to a wavelength scanning and detector device.

Recently developed spectrophotometers are based on dielectric filters and an array of semiconductor detectors containing no mechanical, moving parts (Figure 2). A linear array of, e.g., 1024 closely packed individual photodiodes is covered by a wedge filter, which allows the passage of a specific narrow wavelength range to each underlying photodiode. The skin is irradiated by a short quasicontinuous xenon flash lamp, and the reflected light is recorded by the wedge filter-covered photodiode array. The wavelength specificity is highly dependent on the quality of the light filter and the number of individual photodiodes comprising the array. These systems are developed for industrial use in the painting industry and generally show intermeasurement variation far below 1% *in vitro* allowing several measurements in the entire spectrum in seconds. This precision generally exceeds the need in most biological systems.[1] However, depending on the design of the detecting devise some loss of accuracy may be induced in *in vivo* experiments. In most investigations the vascular status is of importance making it essential to standardize and reduce the weight of the measuring equipment onto the skin surface.[1,4,5,15] Especially in units with xenon flash systems built into the measuring head, large physical size may impede accurate repositioning onto the measuring site (Figure 2). By small adaptation for *in vivo* measurements most commercial systems will be able to meet these requirements, but it is very important to describe the nature of the technical alternations exactly to allow data comparisons between laboratories.

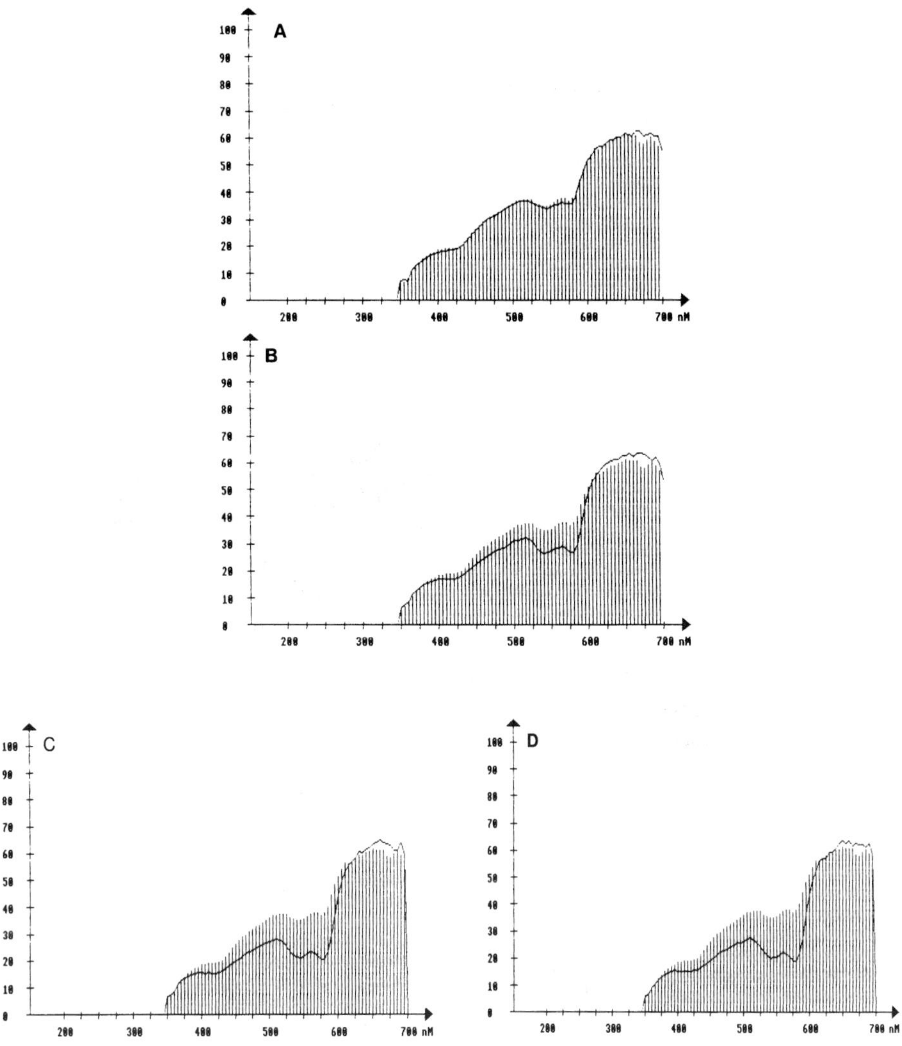

FIGURE 1   Reflectance measurement on normal skin. The figures shows a series of measurements on adjacent normal skin (horizontal lines) and UVB-induced erythema (solid line) with increased UVB exposure from A to D giving relative reflectance (*y*-axis) as a function of wavelengths (*x*-axis). Generally skin reflectance increases from the near ultraviolet range and throughout the visible part of the light spectrum (see text for further details).

III. SKIN OPTICS

In the skin several major absorbing molecules coexist in a heterogeneous way making the optical properties of skin very complicated.[3,16] Furthermore, broad and multiple absorption bands of skin chromophores and diffuse scattering of the incoming light make it complex optically to define the events.[3,11,16,17]

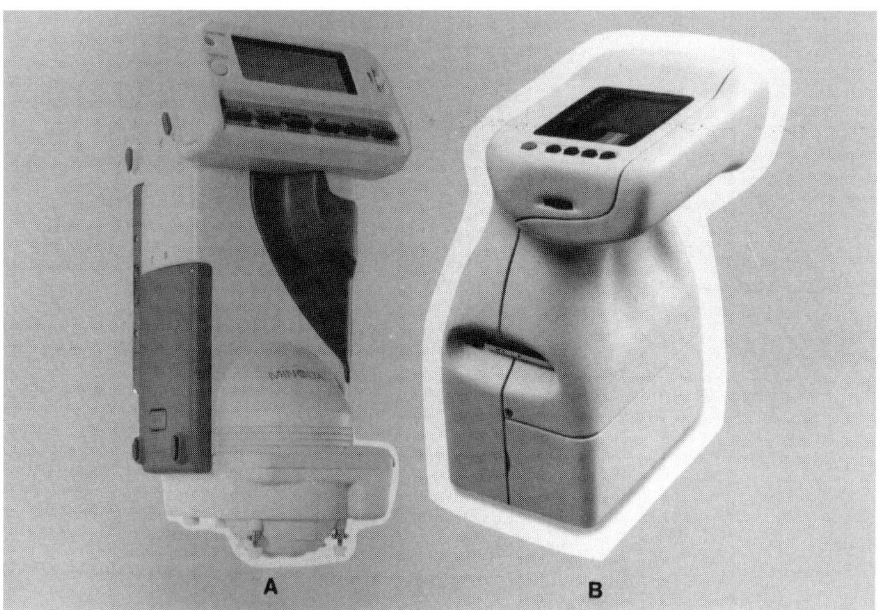

FIGURE 2 The figure shows the two different reflectance spectroscopic units based on advanced techniques both using dielectric filters and semiconductor arrays. Both the Minolta system (A) and the Ziess equipment (B) are handheld units and very accurate. Both companies have several units. These two units are presented without any special recommendation.

When skin surface is irradiated with light, entering photons may be both absorbed or scattered by different cutaneous molecules.

In vivo the stratum corneum predominantly induces diffuse forward scattering and influence on the total skin reflectance is minimal.[3,16] The epidermal melanin heavily absorbs all wavelengths, but is strongest attenuating the near ultraviolet range of the spectrum.[3,16,18] Both oxygenized and deoxygenized hemoglobin absorbs the incoming light specifically and influence *in vivo* reflectance spectrograms accordingly.[1,3,16] Oxygenized hemoglobin absorbs strongest at 405 nm with a relative low absorbance from 430 nm until two characteristic absorbance maxima at 550 and 575 nm give a double peak appearance (Figure 3).[16] Deoxygenized hemoglobin absorbs strongest at 430 nm followed by relative low attenuation until 550 nm (Figure 3).[16] Both hemoglobins absorbs minimally from 620 nm and throughout the spectrum (Figure 3).[16]

Although the hemoglobins absorb potently from 405 until 430 nm *in vivo* hemoglobin is most easily analyzed from 550 until 575 nm (Figure 1).[1,3,11] This is partially due to the strong melanin influence in the near ultraviolet range, but is also caused by cutaneous scattering properties of especially the epidermis allowing diminished penetration of shorter wavelengths.[16,18] Furthermore, dermal collagen, which purified appears almost 100% white,[19] will reflect all nonabsorbed wavelengths, but since shorter wavelengths more easily scatter, the longer wavelengths tend to penetrate deeper into the dermis, and thus more

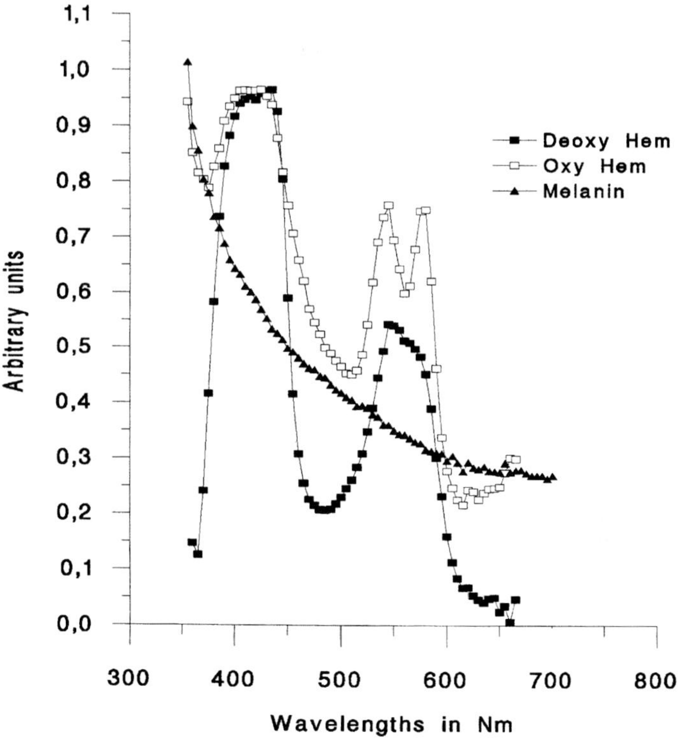

FIGURE 3 Chromophore *in vitro* absorbance. As a function of wavelength *in vitro* absorbance of melanin and oxygenized and deoxygenized hemoglobin is shown. Melanin has the strongest absorption in the near UV-range with a gradually diminished attenuation. The two forms of hemoglobins absorb specifically (see text for details).

likely are absorbed by hemoglobins. Therefore, light of shorter wavelengths are less well represented in the papillary dermis and hence sparsely absorbed by hemoglobin molecules. *In vivo* skin reflectance gradually increases from the ultraviolet range, until the near infrared part of the spectrum caused by *in vivo* melanin with minor hemoglobin influence from 405 to 430 nm and more pronounced at effects seen from 550 to 575 nm (Figure 1).[1]

IV. DATA ANALYSIS TECHNIQUES

To perform chromophore comparisons several authors have suggested a simplified three-layer model allowing the introduction of pigment indexes as a measure for chromophore content in the skin.[3,11,17,20] Optically, the human skin is then described in terms of two different heavily absorbing and only mildly scattering layers on top of a nearly absolute reflecting collagen layer in the dermis (Figure 4).[3,11,17,20] The outer stratum corneum, which mainly induces diffuse forward scattering, allows penetration of all wavelengths into the

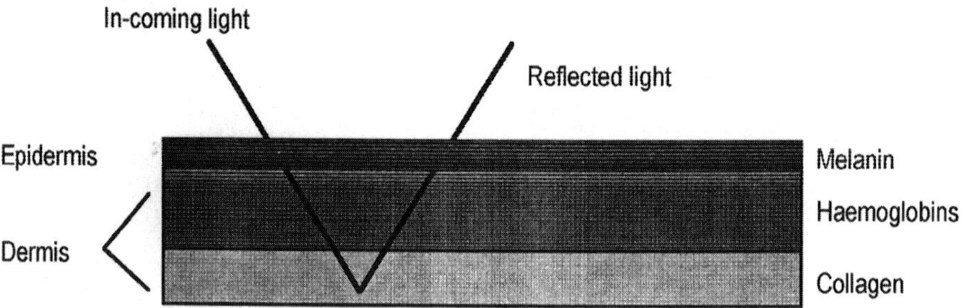

FIGURE 4   Using the simplified three-layer skin model it is assumed that incoming light first reaches the epidermal melanin. The nonabsorbed part, which penetrates the epidermis, then reaches the hemoglobin-containing papillary dermis. Finally, the light reflected by dermal collagen reaches the skin surface passing the hemoglobins and melanin-containing layers a second time.

stratum corneum. The incoming light is here intensely absorbed by the epidermal melanin and the nonabsorbed part, which penetrate the epidermis and reach the hemoglobin containing papillary dermis.[3,11,16,18,21] This layer has the most pronounced dermal influence on skin color and the remaining part of the light is then diffusely reflected by dermal collagen almost linearly reaching the skin surface passing the hemoglobin and melanin-containing layers twice.[3,16,22]

A.   PIGMENT INDEXES

Using very simple mathematical equations several groups have developed very useful pigment indexes calculating the reflectance in specific parts of the spectrum to estimate cutaneous chromophore content. It is assumed that "optical windows" exist allowing different chromophore amounts to be calculated by mathematical analysis of specific and separate parts of the measured *in vivo* skin reflectance. Pigment indexes were proposed both to describe melanin pigmentation, bilirubin amounts, and the degree of erythema.[1,3,11,18,23] Due to the strong UV absorbance of melanin it has been proposed that the slope of the *in vivo* reflectance spectrogram at 365 to 395 nm correlates with melanin content in the skin.[1,16] Some authors have also suggested using the integrated slope from 705 to 800 nm to measure melanin pigmentation.[16,18,21] Theoretically, it is most reasonable to determine cutaneous melanin amounts in the range of the spectrum, where it is the major absorbing molecule.[16] However, some hardware systems have low reproducibility in the ultraviolet range and since hemoglobins have very low absorbance in the near infrared range this region of the spectrum may serve as an alternative range to calculate melanin amounts with minimal hemoglobin influence.[16,18] Most groups, however, agree on determining erythema in the area of the spectrum around the double peak from oxygenized hemoglobin. The most widely used "erythema index" was originally defined by Dawson et al.[3] Since none of the skin chromophores absorbs in narrow bands, both high melanin pigmentation and severe erythema could influence the calculations of other chromophore contents.[1,16]

B. MULTIPLE REGRESSION ANALYSIS OF CHROMOPHORE CONTENTS

To improve data analysis further and allow transcutaneous chromophore concentration calculations the following assumptions were made. Since absorption by melanin and hemoglobins was much more prominent than the scattering induced by these layers, the transmittance can be approximated according to lambert Beers law.[3,11,16] Several groups have suggested different analysis techniques to estimate pigment contents using multiple regression analysis, but in our hands the Choleski decomposition method worked best.[24,25] By this mathematical method the optical amount of the skin chromophores can be calculated, if the wavelength-dependent chromophore absorption coefficients are determined.[24] Essentially, the Choleski decomposition method of multiple regression analysis allows the use of n chromophores and the model can be extended to cover bilirubin or exogene-induced pigments. After measuring *in vivo* skin reflectance chromophore amounts are detected using only parts of the spectrum and subsequently incorporated into the calculations of the following pigments. *In vivo* melanin amounts are calculated from 360 to 390 nm, which corresponds to its main absorbing area in the recorded spectrum and also the area with minimal absorbance of other chromophores.[20] Hemoglobins are then analyzed from the 515 to 610 nm band incorporating the previously determined melanin amount.[20] This means that the degree of erythema is calculated after the computer has "depigmented" the skin to reduce chromophore covariation to a minimum. For most practical purposes other chromophores are included only if these pigments are present in relevant amounts.[20]

V. REFLECTANCE SPECTROSCOPY IN DERMATOLOGICAL RESEARCH

Often authors compare sensitivity of different types of equipment when several bioengineering techniques are included to describe the same biological phenomena.[26,27] For example, erythema of the skin can be assessed by scanning reflectance spectroscopy, tristimulus color measurements, laser Doppler blood flowmetry, and temperature measurement. It is important to realize that although the different types of measurements may directly or indirectly correlate to identical physiological values, vast differences exist between related measuring techniques.[1,4,16,17] Tristimulus systems using the CIE three-dimensional color coordinate system measures skin brightness and utilizes two color coordinates to express green/red and blue/yellow relations.[28] Methods based on red-green photodiodes use the logarithmic relation between green and red reflectance to describe skin erythema and red reflectance to measure melanin pigmentation.[4] Since reflectance spectroscopy measures skin reflectance within a continuous spectrum, differences between various techniques may be observed, e.g., when erythema or vasoconstriction is investigated using reflectance spectroscopic measurements or *laser Doppler flowmetry*. Although laser Doppler flowmetry and reflectance spectroscopy often show good correlation, it is important to realize the differences between the two measured optical phenomena.[2,3] As a measure of cutaneous blood *flow* laser Doppler flowmetry utilizes a monochromatic high intensity laser beam directed onto the skin surface to determine the Doppler shift caused by moving erythrocytes.[2] Therefore, laser Doppler flowmetry includes only *moving* hemoglobin, but reflectance spectroscopy measures both arterial and venous hemoglobin absorbances independently of cutaneous blood flow.[3] Since laser Doppler flowmetry can be performed at different wavelengths with moderate to high tissue penetration depths these measurements may or may not be influenced by the blood flow of the deeper dermal vessels, which are normally not measured using scanning reflectance spectroscopy.[2,15] Specificity and sensitivity of laser Doppler flowmetry and reflectance spectroscopy are dependent on the nature of the vascular phenomena in questing and probably seldom on different equipment sensitivity.

Today *in vivo* reflectance spectroscopic measurements allow exact analysis of skin pigmentation and skin color changes. In several publications the method has proven both sensitive and reliable; dermatological models and treatment modalities have been investigated giving new information about vascular and pigment changes.[10,29-32] To investigate cutaneous vascular changes we have especially enjoyed the combination of laser Doppler flowmetry and reflectance spectroscopy. In particular, we have concentrated on UVB erythema, skin irritation, and corticosteroid vasoconstriction.[15,29,30,33,34] During these experiments new information was achieved, especially on cutaneous vascular reactions. The combined use of multiple regression data analysis, which allows erythema to be expressed as venous or arterial, and laser Doppler measurements gave reliable information on both cutaneous blood flow and hemoglobin amounts. Generally almost any cutaneous condition associated with skin color changes is suited for reflectance spectroscopic investigation. Using objective measurements it is possible to quantitate pigment changes and increase our knowledge of *in vivo* cutaneous biophysical phemonena. However, to increase the use of this informative technique it may be recommended that systems especially designed for cutaneous investigations be developed.

ACKNOWLEDGMENTS

I would like to thank K. Abrams, P. Broichmann, K. Kubota, K. Milioni, and H.I. Maibach from the Department of Dermatology, University of California, San Francisco.

References

1. **Bjerring, P. and Andersen, P.H.,** Skin reflectance spectrophotometry, *Photodermatology*, 4, 167, 1987.
2. **Nilsson, G.E., Tenland, T., and Oberg, P.,** Evaluation of a laser Doppler flowmeter for measurement of tissue blood flow, *IEEE Trans. Biomed. Eng.*, 27, 597, 1980.
3. **Dawson, J.B., Barker, D.J., Ellis, D.J., Grassam, E., et al.,** A theoretical and experimental study of light absorption and scattering by in vivo skin, *Phys. Med. Biol.*, 25, 695, 1980.
4. **Polla, L.L., Margolis, R.J., Dover, J.S., Whitaker, D., et al.,** Melanosomes are a primary target of Q-switched ruby laser irradiation in guinea pig skin, *J. Invest. Dermatol.* 89, 281, 1987.
5. **Diffey, B.L., Oliver, R.J., and Farr, P.M.,** A portable instrument for quantifying erythema induced by ultraviolet radiation, *Br. J. Dermatol.*, 111, 663, 1984.
6. **Frank, K.H., Kessler, M., Appelbaum, K., and Dummler, W.,** The Erlangen micro-lightguide spectrophotometer EMPHO I, *Phys. Med. Biol.*, 34, 1883, 1989.
7. **Nilson, G.,** Measurement of water exchange through skin, *Med. Biol. Eng. Comput.*, 15, 209, 1987.
8. **Edwards, E.A. and Duntley, S.Q.,** The pigments and color of human living skin, *Am. J. Anat.*, 65, 1, 1939.
9. **Feather, J.W., Ryatt, K.S., Dawson, J.B., Cotterill, J.A., et al.** Reflectance spectrophotometric quantification of skin colour changes induced by topical corticosteroid preparations, *Br. J. Dermatol.*, 106, 437, 1982.
10. **Ryatt, K.S., Feather, J.W., Dawson, J.B., and Cotterill, J.A.,** The usefulness of reflectance spectrophotometric measurements during psoralens and ultraviolet A therapy for psoriasis, *J. Am. Acad. Dermatol.*, 9, 558, 1983.
11. **Wan, S., Jaenicke, K.F., and Parrish, J.A.,** Comparison of the erythemogenic effectiveness of ultraviolet-B (290–320 nm) and ultraviolet-A (320–400 nm) radiation by skin reflectance, *Photochem. Photobiol.*, 37, 547, 1983.
12. **Jacquez, J.A. and Kuppenheim, H.F.,** Spectral reflectance of human skin in the region 235–1000 nm, *J. Appl. Physiol.*, 7, 523, 1955.
13. **Jacquez, J.J., Kuppenheim, H.F., and Dimitroff, J.M.,** Spectral reflectance of human skin in the region 235-700 mm, *J. Appl. Physiol.*, 8, 212, 1955.
14. **Jacquez, J.A., Wayne, J.H., McKeehan, W., Dimitroff, J.M., et al.,** Spectral reflectance of human skin in the region 0.7–2.6 m, *J. Appl. Physiol.*, 8, 297, 1955.
15. **Andersen, P.H., Abrams, K., Bjerring, P., and Maibach, H.A.,** Time-correlation study of ultraviolet B-induced erythema measured by reflectance spectroscopy and laser Doppler flowmetry, *Photodermatol. Photoimmunol. Photomed.*, 8, 123, 1991.
16. **Anderson, R.R., Parrish, J.A., and Jaenicke, K.F.,** Optical properties of human skin, in *The Science Photomedicine,* Reagan, J.D. and Parrish, J.A., Eds., Plenum Press, New York, 1982, 147.

17. **Wan, S., Parrish, J.A., and Jaenicke, K.F.**, Quantitative evaluation of ultraviolet induced erythema, *Photochem. Photobiol.*, 37, 643, 1983.

18. **Kollias, N. and Baqer, A.**, Spectroscopic characteristics of human melanin in vivo, *J. Invest, Dermatol.*, 1985.

19. **Andersen, P.H. and Bjerring, P.**, Spectral reflectance of human skin in vivo, *Photodermatol. Photoimmunol. Photomed.*, 7, 5, 1990.

20. **Andersen, P.H. and Bjerring, P.**, Noninvasive computerized analysis of skin chromophores in vivo by reflectance spectroscopy, *Photodermatol. Photoimmunol. Photomed.*, 7, 249, 1990.

21. **Kollias, N. and Baqer, A.H.**, Quantitative assessment of UV-induced pigmentation and erythema, *Photodermatology*, 5, 53, 1988.

22. **Findlay, G.H.**, Blue skin, *Br. J. Dermatol.*, 83, 127, 1970.

23. **Strange, M. and Cassady, G.**, Neonatal transcutaneous bilirubinometry, *Clin. Perinatol.*, 12, 51, 1985.

24. **Makridakis, S.G.**, Multiple regression, in *Forecasting, Methods and Applications*, Madrikakis, S.G, Ed., John Wiley, New York, 1983, 246.

25. **Martin, R.S. and Peters, G.**, Symmetric decomposition of a positive definity matrix, *Numer. Math.*, 7, 362, 1965.

26. **Agner, T.**, Noninvasive measuring methods for the investigation of irritant patch test reactions. A study of patients with hand eczema, atopic dermatitis and controls, *Acta Dermatol. Venereol. Suppl. Stockh.*, 173, 1, 1992.

27. **Agner, T. and Serup, J.**, Skin reactions to irritants assessed by non-invasive bioengineering methods, *Contact Derm.*, 20, 352, 1989.

28. **Robertson, A.R.**, The CIE 1976 color difference formulas, *Color Res.* 7, 643, 1977.

29. **Andersen, P.H., Milioni, K., and Maibach, H.**, The cutaneous corticosteroid vasoconstriction assay: a reflectance spectroscopic and laser Doppler flowmetric study, *Br. J. Dermatol.*, 128, 660, 1993.

30. **Andersen, P.H., Abrams, K., and Maibach, H.**, Ultraviolet B dose-dependent inflammation in humans: a reflectance spectroscopic and laser Doppler flowmetric study using topical pharmacologic antagonists on irradiated skin, *Photodermatol. Photoimmunol. Photomed.*, 9, 17, 1992.

31. **Tang, S. and Gilchrest, B.**, Spectrophotometric analysis of normal, lesional and treated skin of patients with port wine stains (PWS), *J. Invest. Dermatol.*, 78, 340, 1982.

32. **Queille-Roussel, C., Poncet, M., and Schafer, H.**, Quantification of skin colour changes induced by topical corticosteroid preparations using Minolta Chromameter, *Br. J., Dermatol.*, 124, 264, 1991.

33. **Andersen, P.H., Bucher, A.P., Saeed, I., Lee, P.C., et al.**, Faecal enzymes: i vivo human skin irritation, *Contact Derm.*, 30, 152, 1994

34 **Andersen, P.H., Broichmann, P.W., and Maibach, H.**, A corticosteroid, non-steroidal anti-inflammatory drug and an antihistamine modulate in viv vascular reactions before and during post-occlusive hyperaemia, *Br. J. Dermatol.*, 128, 137, 1993.

6 Mouse Models for Scaly Skin Diseases

John P. Sundberg, Harm HogenEsch, and Lloyd E. King, Jr.

TABLE OF CONTENTS

I.   INTRODUCTION

Although mouse mutations have many advantages as animal models, which have been described elsewhere,[1-3] there are also many disadvantages that mainly relate to structural or functional species differences. Failure to recognize these differences may result in misinterpretation of the biological processes being studied or dismissal of the model because it fails to meet a set of preconceived notions on what one expects of an animal model.

Limitations of mice as a species for use as animal models for human dermatological diseases begin with the fact that most strains of mice are covered with hair. The density of follicles is much higher than in humans making it harder to process for biochemical and microscopic studies. Mouse hair follicles produce a single hair shaft and are not compound. The fur or pelage hairs have four distinct types, which are different from humans. In addition, there are a number of hair types not found on humans at all, such as vibrissae and tail hairs.[4] The hair of mice cycles in a wave pattern from the head to the tail[5,6] as opposed to a mosaic pattern as in humans.[7] Most adult mouse follicles are in telogen for prolonged periods of time, explaining why hair length remains constant, as opposed to humans where it constantly grows.[8]

Genetic control of pigmentation or the lack of pigmentation of mouse hairs is complex, but has been studied more extensively than in humans.[9,10] In contrast to humans, the interfollicular epidermis of mice is rarely pigmented.[11] Pigment is found in the bulb of actively growing mouse follicles, which enables one to identify the new growth associated with hair cycles or induced by trauma in shaved or depilated mice by the line or site of pigmentation. Dermal melanophages indicative of pigmentary incontinence are therefore commonly observed.

It is visually difficult to quantitate inflammation in the skin of mice as compared to human skin except by measuring increased thickness due to dermal edema. When the fur is removed by shaving or depilatories, the underlying skin appears to be pink or inflamed since mouse skin is very thin. This pink or reddish color is due to the underlying, well vascularized tissues that become readily evident when the hair is removed. Some mutations, such as flaky skin (*fsn*), have a marked dermal inflammatory cell component with dilated dermal capillaries that causes the skin to appear red. However, *fsn/fsn* mice are also severely anemic, which complicates this factor of clinical interpretation.

Several specific differences exist between mouse and human skin. One is that mice lack apocrine sweat glands.[8] Mammary glands, a modified form of apocrine sweat glands, are located beneath almost all of the truncal skin in the subcutaneous fat.[12] A second difference is that newborn mouse epidermis is thick when pups are first born but becomes quite thin within the first 2 weeks of life.[13] A relatively thin epidermis is usually present in mice with a normal coat of hair or fur. A thickened epidermis is almost always observed when alopecia or a thinned hair coat is present in mice with mutations in which alopecia is a major phenotype.[11,14]

A number of mouse mutations exist and are readily available for study of skin diseases.[15-17] However, unlike human scaly skin diseases, scaly mouse skin can be difficult to differentiate at the gross level primarily because most of the body is covered with hair. Variations in anatomic sites (flaky tail [tail] vs. flaky skin [torso]) may assist in this process. Clinical and histological features of a single gene mutation may vary in inbred mice depending upon the strain the mutation is maintained on. In a similar manner, clinical variability can be observed within human families with the same inherited defect due to variations in other genes between individuals. For example, *fsn* has essentially no stratum granulosum on the A/J inbred background compared to a very prominent layer on the BALB/cByJ background.[18]

Alopecia may or may not be a feature of the mutation. Some scaly and alopecic skin mutations respond dramatically to topical irritants (flaky skin; Sundberg, unpublished observations) while others do not (BMP–4).[19] Epigenetic effects, such as the presence of ectoparasites (mites) or environment (low humidity associated with chronic ulcerative dermatitis), can produce scaly skin lesions with alopecia in normal mice if they are not housed properly.[20,21]

Just as it may be hard to determine the underlying defect(s) in "scaly" mice, deciding whether a "flaky" human skin disease is a form of ichthyosis or psoriasis may not be so simple. These difficulties are markedly increased when expanding these concepts to include skin conditions that grossly or microscopically look like ichthyosis (ichthyosiform) or psoriasis (psoriasiform). Conceptually, these two terms are often used to separate a congential, noninflammatory, symmetrically located, noncorticosteroid responsive, scaly skin disease (ichthyosiform) from a red, scaly, sometimes pustular, but not contagious, inflamed skin condition that is often corticosteroid responsive, and induced asymmetrically by trauma (psoriasiform). Pathologically, these conditions are further identified by ensuring that the flaky skin disease is not due to pathogens and has a thickened, nonviable outer epidermal layer, the stratum corneum. Ichthyosiform conditions usually do not have any inflammatory cells (no neutrophils or lymphocytes), a variable granular cell layer (thin, thick, unchanged), and a variable viable cell (Malpighian) layer that may or may not be thickened or have increased proliferation. In contrast, psoriasiform diseases often have increased epidermal thickness to account for the plaques palpated clinically, neutrophils in the outer epidermal layer (stratum corneum), and elongated capillaries in the dermal papilla. A number of very common noninfectious chronic skin diseases with known and unknown causes (such as allergic contact dermatitis, chronic scratching, or eczema) can produce an almost identical histological pattern. Because of the common use of skin biopsies to confirm the diagnosis of a chronic red, scaly rash and the inability to separate psoriasis from its histological and clinical imitators, the term "psoriasiform dermatitis" has become a commonly accepted term. It is often overlooked that this term, "psoriasiform dermatitis," has differing meanings to clinicians, pathologists, and investigative scientists. Before deciding which, if any, of the mouse models described below are likely to be useful, the fundamental questions of which aspect of an ichthyosiform or psoriasiform human skin disease is to be modeled should be carefully defined.

II. MOUSE MUTATIONS WITH SCALY SKIN PHENOTYPES

Mouse mutations arise spontaneously or can be induced. Induced mutations have been created consistently in the past using various forms of radiation or chemical mutagens, such as ethyl nitrosourea. More recently, transgenic and homologous recombination (gene knockout) techniques have been developed to successfully generate numerous mouse mutations. The resulting mutations are usually recessive. Gene symbols are used to designate the specific mutation and are usually a short abbreviation of the descriptive name (unless the specific mutated protein product is known). The symbol is always italicized. Recessive mutations have symbols that are written in lower case. Dominant and semi-dominant mutations use symbols that begin in upper case letters.

For these mutations to be useful as models, they must be readily available to the research community, such as through the Mouse Mutant Resource or Induced (transgenic and knockout) Mutant Resource at The Jackson Laboratory. Furthermore, to ensure uniformity of results within a study or between institutions by scientists working independently, the mutation should be on an inbred rather than a hybrid or outbred genetic background, if possible. The mutations reviewed below meet these two critical criteria or will so in the near future.

A.   ASEBIA (GENE SYMBOL: *ab*)

The asebia (symbol: *ab*) mouse mutation arose spontaneously in a colony of BALB/cCrglGa mice. Asebia is due to an autosomal recessive gene with complete penetrance. Mutant mice were inbred and maintained on a background designated BALB/cGa +/*ab*.[22] A similar mutation, allelic with asebia, arose spontaneously in the BALB/cJ inbred strain at The Jackson Laboratory. The remutation, designated *ab^J*, was mapped to chromosome 19.[23] Asebia[J] is maintained as an inbred strain ABJ/Le.[24]

Homozygous *ab* or *ab^J* mice exhibit abnormalities of hair growth that can sometimes be detected as early as 7 days of age. Alopecia increases in severity with successive hair cycles (Figure 1). Fine epidermal scaling may be mild but becomes more severe with age. Mutant mice are photophobic, exhibit pruritus of the eyelids, and develop a sticky exudate that encrusts the eyelids (Figure 2). Mutants are smaller than heterozygous controls.[21,25]

FIGURE 1   Two ABJ/Le littermates. The left mouse is normal (+/*ab^J*) and the right mouse has alopecia typical of the asebia mutation (*ab^J*/*ab^J*).

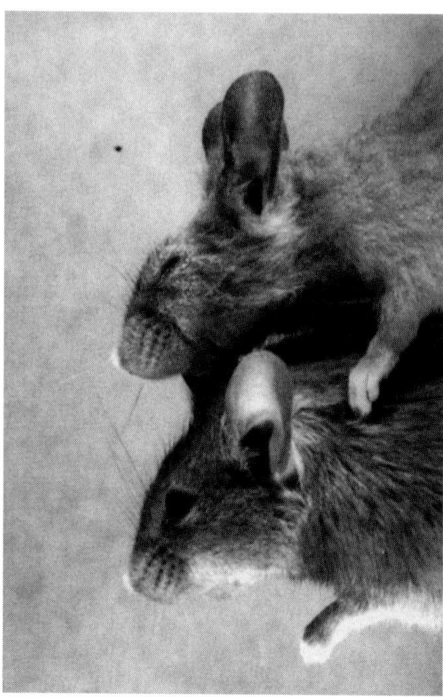

FIGURE 2 Photophobia in the asebia mutation (top) is associated with Meibomian gland hypoplasia.

The original description reported an absolute lack of sebaceous glands in *ab* mice.[22] Subsequently several groups confirmed that both *ab* and *ab^J* mutations have hypoplastic, not aplastic, sebaceous glands (Figures 3 and 4).[25-27] Sebaceous glands lack an orderly maturation of peripheral reserve cells into vacuolated holocrine cells that rupture to secrete their contents into the sebaceous duct. Consequently, the glands are bud-like in appearance, very small, and easily overlooked (Figures 5 and 6). Sebaceous glands and the modified sebaceous glands of the eyelids (Meibomian glands) appear to be disorganized with reserve cells interspersed among sebaceous cells in various stages of differentiation.[25,26] Other glands, including preputial, clitoral, anal glands, and eccrine glands of the footpads, are not affected.[25,28]

Mutant mice also exhibit alterations of their epidermis,[22,29] dermis,[29,30] and hair follicles.[31] The epidermis becomes moderately orthokeratotic, which increases in severity with age.[26,29] The dermis is thicker and more vascular than littermate controls with various degrees of inflammation.[29] The inflammation is due to the presence of numerous mast cells and macrophages that contain both intact and disintegrating material considered to be lipid.[30,32] The hair cycle is delayed in this mutation and mutant hair follicles extend deep into the hypodermis compared to controls.[26,31] Defects of the inner and outer follicular root sheaths have been proposed as the problem with hair development.[31]

FIGURE 3   Tail hair follicles have essentially no sebaceous gland in asebia (*ab^J/ab^J*) mice.

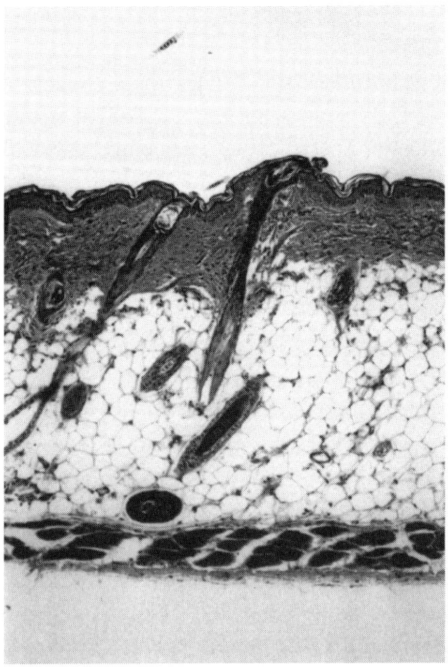

FIGURE 4   Truncal epidermis is mildly acanthotic. Follicles have rudimentary sebaceous glands in asebia (*ab^J/ab^J*) mice.

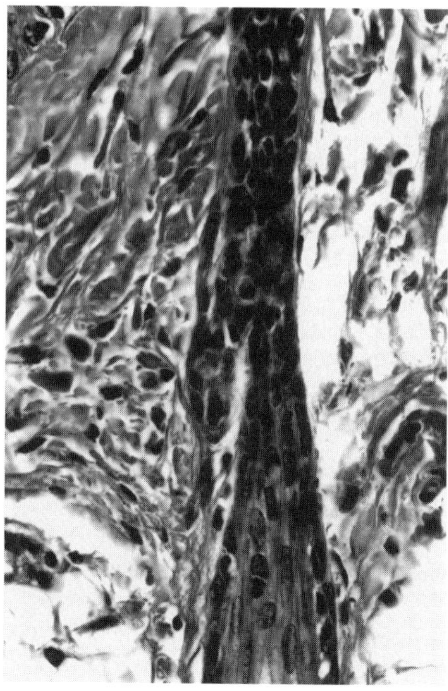

FIGURE 5 Higher magnification of Figure 4 to illustrate the rudimentary sebaceous gland.

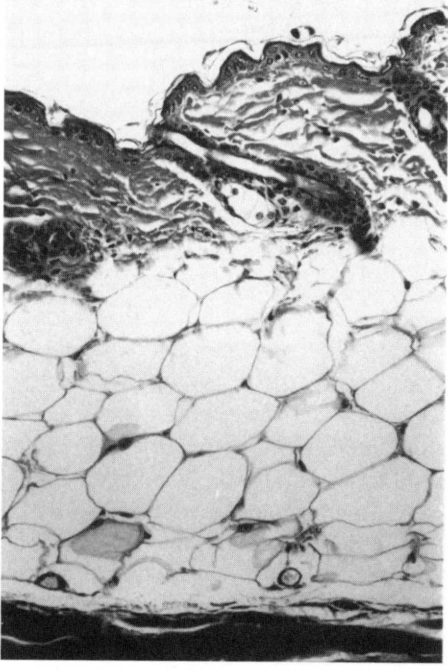

FIGURE 6 Dorsal skin from a littermate control ($+/ab^J$) mouse illustrating normal epidermal thickness and sebaceous gland morphology in an adult.

Plucked and mounted hairs from asebia mice had few, if any, awl hairs compared to littermate controls.[28] Examination by scanning electron microscopy of plucked hairs from affected mice, revealed various amounts of surface debris along the length of the hair shaft, but no apparent changes in basic morphology from normal littermates.[28]

Immunologic abnormalities have not been described, however, *ab* mice are more susceptible to infantile diarrhea than controls,[22] suggesting that this should be investigated. Skin graft studies have established that the *ab^J* gene activity in the epidermis was responsible for sebaceous gland abnormalities, but a distant site was responsible for changes in hair cycle times.[26] Limited studies of lipids found in the skin of mice indicated that *ab* mice were deficient in sterols esterified with very long-chain fatty acids, wax esters, and wax diesters, but the mutant skin was rich in free sterols.[33]

Kinetic studies were done to determine circadian rhythms in cell proliferation of chronically hyperproliferative epidermis of *ab/ab* mice. It was found that the circadian rhythm was depressed in comparison to BALB/cJ epidermis and that of strains used in ten other published studies.[34]

A number of dermatologists have commented, based on the epidermal hyperplasia, increased dermal vascularity, and inflammation, that *ab* or *ab^J* mice might be useful models for the human disease psoriasis.[30] However, this has been discounted by others.[25,35,36]

The ABJ/Le-*ab^J* inbred strain is preserved in The Jackson Laboratory Frozen Embryo Repository. Embyros can be thawed, implanted into pseudopregnant females, and live mice born with the mutation when needed for research. DNA extracts are available from the Mouse DNA Resource at The Jackson Laboratory.

B.   BARE-PATCHES (GENE SYMBOL: *Bpa*)

Bare-patches is a radiation-induced mutation that arose in the (C3H/HeJ × 101/H)F1 progeny of a male mouse.[37,38] Heterozygous females (*Bpa/+*) can be identified around 5 days of age by foci of alopecia at the time the first hairs are emerging. Flakes and scabs develop around 8 days of age and persist for weeks in these alopecic foci. The alopecic foci overlap areas with normal hair by 12 days of age.[38] Adult heterozygous females can be recognized by the presence of narrow, transverse stripes due to the dark base of immediately posterior hair follicles being evident in alopecic areas.[16] *Bpa/+* females are smaller than siblings and may have abnormal feet. Toes of the rear feet may be short and bent. The ears and tail may develop abnormally, presumably due to formation of scabs. Hemizygous males die before birth.[38] Slit lamp studies revealed that 8 of 10 *Bpa/+* mice developed cataracts. All of the +/+ mice examined had normal lenses.[39]

Bpa/+ female mice develop a hyperkeratotic eruption around day 8 of life that lasts several weeks. The hyperkeratosis appears in a linear and blotchy pattern of a nonbullous ichthyosis type that extends into hair follicles.[39] Linear hyperpigmentation occurs in *Bpa/+* mice.[39]

Von Kossa stains of skeletal sections can be used to reveal stippled foci of premature mineralization within columns of chondroblasts of vertebrae and long bones. This feature is absent in sections of control mouse bones.[39]

Bare-patches is a semidominant, sex-linked, male-lethal gene. It was originally suggested that *Bpa* was an allele of tabby *(Ta)*,[38] striated *(Str)*, or harlequin *(Hq)*;[38,40] however, genetic mapping places all four of these mutations at different locations on the X Chromosome. They are now believed to be independent genes. *Bpa* maps very close to the murine X-linked visual pigment gene *(Rsvp)*.[41] Using interspecific backcrosses and analysis of murine yeast artificial chromosomes, the gene for biglycan *(Bgn)* was mapped and was excluded as a candidate gene for the bare patches *(Bpa)* mutation.[42]

Bare-patches is associated with a high production of XO offspring.[38] The genotype was initially called *Fxo*, and was found to be a closely linked factor to *Bpa*.[43] This factor was subsequently demonstrated to be a sex-linked inversion renamed *In(X)1H*.[44,45]

The bare-patches mouse mutation has skeletal, ocular, and cutaneous lesions that are similar to the human genetic disease known as chondrodysplasia punctata.[39,46-48] The human disease is an X-linked dominant mutation characterized by punctate foci of epiphyseal mineralization, cataracts, ichthyosis, and systematized atrophoderma. In both species, the skin lesions are linear and blotchy, reflecting lyonization.[39,46]

The bare-patches mutation was imported to The Jackson Laboratory from the Medical Research Council Radiobiology Unit, Harwell, U.K. in 1976 by Dr. E.M. Eicher. It was crossed to and is maintained on the B6CBACa (C57BL/6J × CBA/CaGnLe) *A*^{w-J}/*A* hybrid background in The Jackson Laboratory Mouse Mutant Resource.

C. CHRONIC PROLIFERATIVE DERMATITIS (GENE SYMBOL: *cpdm*)

Chronic proliferative dermatitis (gene symbol: *cpdm*) was identified in 1991 as a spontaneously occurring disease in the C57BL/KaLwRij strain at the TNO-breeding facility in Rijswijk, the Netherlands. It is an autosomal recessive mutation, based on the incidence of the disease in the offspring of various combinations of breeding pairs.

The skin lesions in *cpdm/cpdm* mice become apparent at 2 weeks after weaning (5 weeks of age) and they are characterized by erythema, partial alopecia, and fine scaling starting in the dorsal neck and ventral chest area (Figure 7).[49] The lesions expand to involve the entire body and head, but ears, footpads, and tail remain unaffected. The animals are severely pruritic.

By light microscopy, the skin lesions are characterized by epidermal hyperplasia with ortho- and parakeratosis, and follicular keratosis (Figures 8 and 9). Individual cell death (apoptosis) of keratinocytes is present to various degrees in the stratum spinosum (Figure 10). There is dilation and proliferation of dermal capillaries. The dermis and epidermis are infiltrated by a mixture of

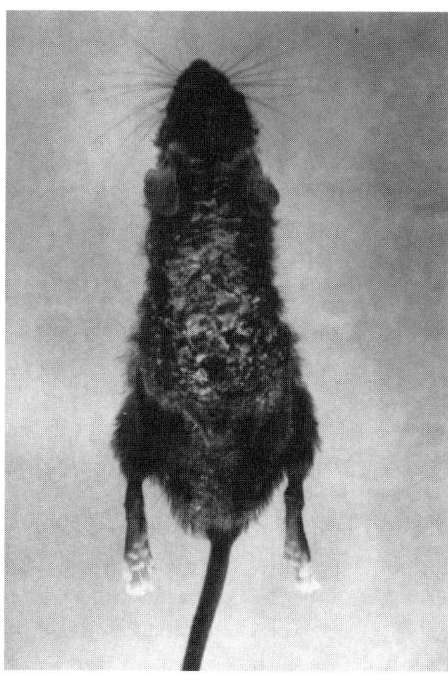

FIGURE 7 Severe scaling is a feature of adult mice homozygous for the chronic proliferative dermatitis mutation (*cpdm/cpdm*).

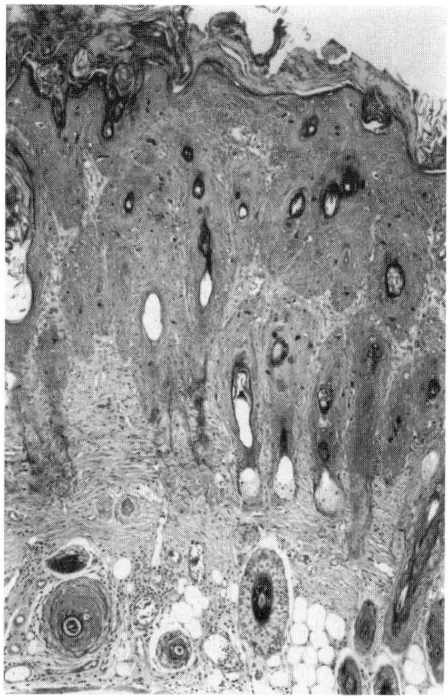

FIGURE 8 Skin from a *cpdm/cpdm* mouse is acanthotic, orthokeratotic, and focally parakeratotic.

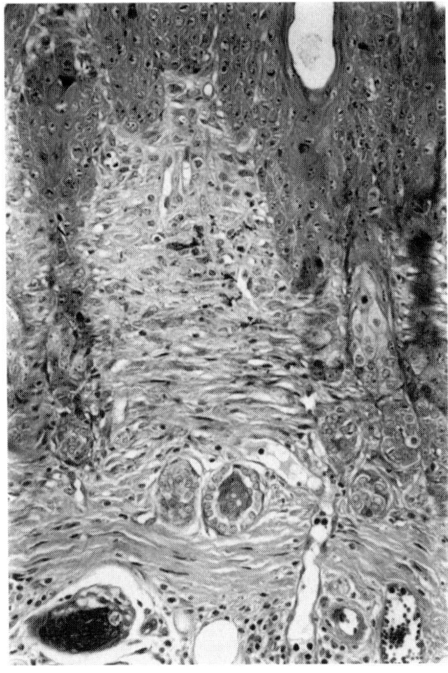

FIGURE 9 Higher magnification of Figure 8 illustrates dermal capillary dilation and sclerosis.

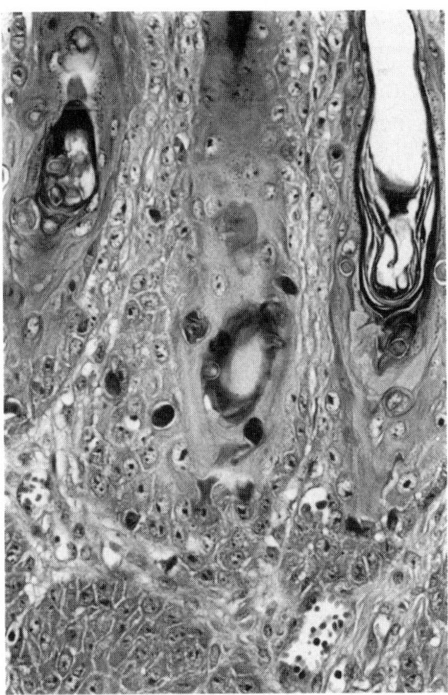

FIGURE 10 Higher magnification of Figure 8 illustrating individual cell necrosis within the follicular root sheaths.

inflammatory cells, predominantly granulocytes and macrophages. Degenerated granulocytes often form microabscesses in the stratum corneum. A considerable proportion of the granulocytes are eosinophils, as confirmed by the presence of granules with a central crystalline bar on electron microscopy. In older mice (>20 weeks of age), the number of granulocytes in the affected skin decreases. Melanin-laden macrophages and dendritic cells are frequently present in the dermis of older *cpdm/cpdm* mice.

The mucosa of the oral cavity, esophagus, and forestomach of the mouse are lined by orthokeratinizing stratified squamous cell epithelium, similar to the epidermis. In *cpdm/cpdm* mice, these organs are afflicted by similar lesions as those in the skin with exception of parakeratosis, which is only observed in the skin. Subcorneal pustules are particularly common in the esophagus, but are also present in the other organs.

Ultrastructurally, apoptotic bodies can be observed, often in the process of being phagocytized by adjacent keratinocytes.[50] Electron-dense spherical inclusions are often present in the mitochondria of keratinocytes in skin lesions. The mitochondria otherwise appear to be intact. These mitochondrial inclusions were never observed in keratinocytes of unaffected skin or other epidermal and dermal cells. Although the significance and pathogenesis of these inclusions are unknown, it is of interest that they are seen in a variety of proliferative skin diseases in mice, including the flaky skin mouse (*fsn/fsn*)[51] and skin lesions induced by topical application of phorbol esters[52] and UV-irradiation.[53]

The hyperproliferative nature of the skin lesions was confirmed by bromodeoxyuridine (BrdU) labeling. An eightfold increase of BrdU-positive keratinocytes in the interfollicular epidermis and a two- to threefold increase of BrdU positive epithelial cells in the esophageal and forestomach epithelium were observed. An increased number of MHC-II positive cells is present in the lesional skin, and most infiltrating cells express Mac-1 (CD11b), consistent with macrophages and granulocytes. Endothelial cells in the skin lesions of *cpdm/cpdm* mice express ICAM-1 (CD54), whereas endothelial cells in the dermis of control mice were negative. An increased expression of ICAM-1 was also seen on basal keratinocytes.[49]

Analysis of the arachidonic acid metabolites prostaglandin E_2 and leukotriene B_4 levels in the skin of *cpdm/cpdm* mice revealed an increase compared to the skin of control mice (Gijbels et al., manucript in preparation). Prostaglandin E_2 and leukotriene B_4 are both potent mediators of inflammation.

Lesions considered to be part of the *cpdm* phenotype affecting other organs include mixed cellular inflammatory lesions in the liver (portal areas and central vein), lung (peribronchial and perivascular), and perisynovial connective tissues of several joints (knee, hip, intervertebral), increased myelopoiesis of the bone marrow, and extramedullary myelopoiesis in the spleen and lymph nodes. In older mice, there is lymphoid depletion of lymph nodes and splenic white pulp.

To determine whether the skin lesions are caused by a local defect or have a systemic component, hematopoietic cell transfer and skin transplantations were performed. Transfer of bone marrow and spleen cells from *cpdm/cpdm* mice to syngeneic control mice did not induce skin lesions in the recipients.[49]

Affected *cpdm/cpdm* skin transplanted onto the skin of syngeneic mice or nude (*nu/nu*) mice maintained its hyperproliferative phenotype over 10 weeks of follow-up, whereas skin of control mice transplanted onto the skin of *cpdm/cpdm* mice remained normal.[54] These results suggest that the initial defect underlying the skin lesions of *cpdm/cpdm* mice is localized in the skin.

Both systemic and topical treatment of the *cpdm/cpdm* skin with cyclosporin A did not ameliorate the lesions. In contrast, subcutaneous injections with corticosteroids (triamcinolone) resulted in a dramatic improvement of the skin condition of the *cpdm/cpdm* mice.[49]

The skin lesions of *cpdm/cpdm* mice include distinct features of psoriasiform dermatitis, such as hyperproliferation of keratinocytes, epidermal infiltration of neutrophils, and dermal capillary dilatation and proliferation. Study of this mouse model and delineation of its pathogenesis should enhance our understanding of mechanisms of proliferative dermatitis in all mammalian species. The chronic nature of the disease makes it a valuable model to evaluate the effect of potential therapeutic compounds on chronic inflammation.

The mice are currently available only through collaboration with researchers in the TNO research institute in the Netherlands, where it is maintained on the C57BL/KaLwRij inbred background. The mice are being imported into The Jackson Laboratory for allelism testing with the flaky skin (*fsn*) mutation and gene mapping. Limited numbers should be available from The Jackson Laboratory within 1 to 2 years.

D. FLAKY SKIN (GENE SYMBOL: *fsn*)

The flaky skin mouse mutation arose as a spontaneous mutation at The Jackson Laboratory in 1985 in the A/J inbred strain.[55] Mutant mice were runted and the gene was poorly penetrant. A/J-*fsn/fsn* mice backcrossed to BALB/cByJ yielded more vigorous offspring, however, they were still difficult to maintain, primarily because of metabolic stress arising from severe anemia. Grafts of *fsn/fsn* ovaries to ovariectomized severe combined immunodeficiency *(scid/scid)* hosts followed by cycles of backcrossing to BALB/cByJ-+/*fsn* males and intercrossing of F1+/*fsn* offspring improved the breeding process. The flaky skin mutation is currently at the N10 generation toward creation of a congenic strain. Flaky skin is an autosomal recessive mutation that maps to the distal end of mouse Chromosome 17.[56]

Mutant (*fsn/fsn*) and littermate controls (+/+ or +/*fsn*, hereafter +/?) can be difficult to differentiate at the gross level until shortly after weaning when the mutants develop thick white scales with associated patchy alopecia. The ventral abdomen develops a fine scale and a banding pattern of thinning hair, reminiscent of embryonic somites. The dorsal skin has a more patchy appearance just after weaning. Skin lesions are progressive, sometimes resulting in a generalized alopecia with age associated thick white scales in severe cases (Figure 11).[36,57,58]

Homozygous (*fsn/fsn*) mice are mildly anemic at birth, a phenotype useful for differentiating mutants from controls. The anemia becomes progressively

FIGURE 11   Thick scales develop on the skin of flaky skin (*fsn/fsn*) mice.

more severe with age. Mutant mice can be identified as early as 1 day of age by generalized pallor or a packed cell volume of less than 40%.[59] As the mutant mice age, their spleens become very large, giving them a "potbelly" appearance. This feature also aids in differentiating mutant mice from littermate controls.[56]

It is difficult to distinguish histologic sections of skin between normal littermate controls and *fsn/fsn* mice during the first week of life. After that point, the thickness of normal skin decreases and the skin rapidly becomes very thin, whereas *fsn/fsn* mouse skin becomes progressively thicker (Figures 12 and 13).[13] The dorsal and ventral skin were histologically similar. The epidermis progressively thickens due to increases in most layers, most notably the stratum corneum. The stratum granulosum was extremely thin in A/J-*fsn/fsn* mice, however, by four to five backcross generations onto BALB/cByJ, this layer became very prominent.[18] The stratum corneum underwent various degrees of orthokeratosis with focal parakeratosis and scattered subcorneal microabscesses. The histological characteristics have been modified by transferring the *fsn* mutation onto a variety of genetic backgrounds.[18] With inbreeding, the focal parakeratosis and microabscesses have become less prominent in certain congenic strains.

In the dermis, capillaries are dilated and there is a mixed inflammatory cell infiltrate. On the original A/J strain and early crosses with BALB/cByJ, neu-

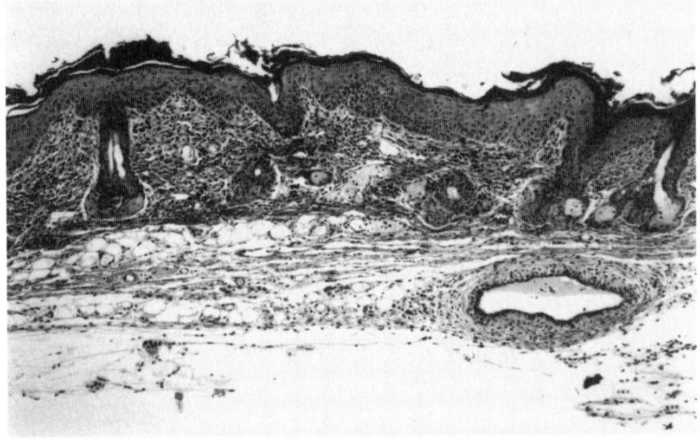

FIGURE 12 Diffuse acanthosis and orthokeratosis with focal parakeratosis and a prominant dermal inflammatory cell infiltrate are the cutaneous features of *fsn/fsn* mice.

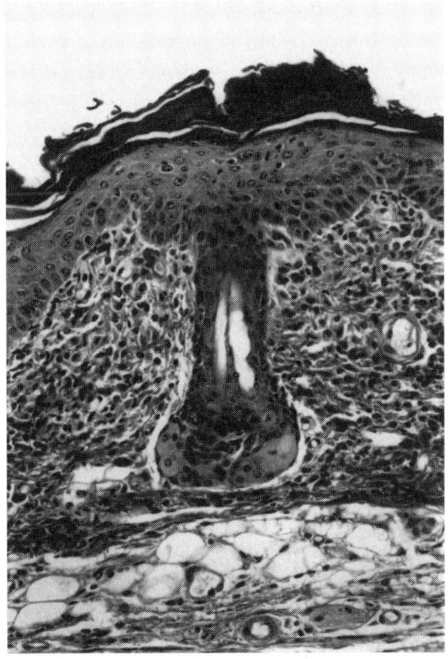

FIGURE 13 Higher magnification of Figure 12 illustrates the focal parakeratosis and dermal cellularity.

trophils comprised a major component of the inflammatory process. This was quantitated by a five- to tenfold increase in myeloperoxidase levels in full thickness biopsies from *fsn/fsn* mice compared to +/? mice.[18] With inbreeding onto the BALB/cByJ background, this has shifted almost exclusively to a mononuclear cell infiltrate. Epidermal hyperplasia has remained, although there has been a concomitant decrease in focal parakeratosis and a change from a thin to nonexistent stratum granulosum (in A/J mice) to a very prominent layer (in BALB/cByJ congenic mice). The neutrophilic response with parakeratosis can be induced by trauma (tape stripping) yielding a positive Koebner reaction.[13,18,36]

Ultrastructural evaluation of skin from *fsn/fsn* and littermate control mice at 42 days of age (N1 and N8 backcross generations) confirmed the marked epidermal hyperplasia with neutrophilic migration into the epidermis of mutant mice. The major alterations included diminished tonofibrils, poor fusion of tonofilaments to keratohyalin granules, intercellular lipid vacuoles in the stratum corneum, and tortuous dermal blood vessels lined by a hyperplastic endothelium with a multilaminated basement membrane. Basal keratinocyte herniations were identified and there were focal areas in which the basement membrane was lost.[51] All of these features have been described in human psoriatic lesions.[60-62] Also found were electron-dense droplets within membrane bound vacuoles in the cytoplasm of keratinocytes and electron-dense mitochondrial inclusions (so called corpus intra cristam),[63] similar to those found in a variety of epidermal hyperproliferative diseases in mice[55] and the chronic proliferative dermatitis mouse mutation described above.[50] However, this interesting feature, observed in mutant mice, has not been described in psoriatic human patients.

Scanning electron microscopy of skin punch biopsies and plucked hairs revealed that the hair shafts had fine cuticular pitting with exophytic projections of various sizes.[51] Clumps of fine cornified debris covered the shafts. Specimens prepared by freeze fracture and examined by scanning electron microscopy clearly demonstrated epidermal thickening with marked orthokeratosis and a loss of hypodermal fat in mutant mice compared to controls.[18,51]

The junction of the glandular and nonglandular portion of the stomach of *fsn/fsn* mice develops a papillomatous lesion that progresses with age. No evidence of a concurrent papillomavirus infection was found.[64]

Scanning electron microscopic evaluation of red blood cells revealed marked variation in size and shape in the *fsn/fsn* mice. There is marked erythropoiesis in both the liver and spleen. Nucleated red blood cells are a major feature of blood cells with vessels in histologic sections.[54] These changes are responses to the severe anemia that develops in the homozygous mice. Changes in diet can eliminate or exaccerbate this condition (Beamer, unpublished data).

Pathologic changes caused by homozygosity for the flaky skin mouse phenotype appears to be determined at the level of bone marrow-derived cells. This hypothesis is based on the observation that bone marrow grafts from *fsn/fsn* mice to *scid/scid* mice induced a skin lesion without other abnormalities, such as anemia, in recipients.[13] This skin lesion does not require the activity of functional lymphoid cells since the skin lesion develops and progresses in mice doubly homozygous (*fsn scid/fsn scid*).[13] The severe combined immunodeficiency (*scid*) mice lack T and B lymphoid cells.[65]

Analysis of eicosanoid levels in mutant skin indicate that 12-hydroxy-5,8,10,14-eicosatetraenoic acid (12-HETE) and leukotriene B_4 are elevated compared to the controls,[18] similar to that described above in the chronic proliferative dermatitis mouse mutation.

Immunohistochemical markers and tritiated thymidine uptake studies have been used to verify changes in protein expression compatible with hyperplasia. The mouse specific keratin, K6, is normally restricted to the inner layer of the outer root sheath of the hair follicle. During hyperplastic responses in *fsn/fsn*, K6 is expressed in the suprabasilar epidermis. Concurrently, K1 and K10, the normal terminal differentiation markers for suprabasilar epidermal keratinocytes, decrease in expression as hyperplasia increases.[66] In normal mouse epidermis, filaggrin is first associated with keratohyalin granules in the stratum granulosum, and later with larger keratohyalin granules in transitional cells, at the junction of the stratum corneum. In lower cornified cells, filaggrin disperses and becomes evenly distributed in the cytoplasm. In the upper cornified cells, filaggrin becomes undetectable. The disappearance of filaggrin in upper cornified cells correlates well with the dispersion of keratin filament bundles, thus providing strong evidence that filaggrin may be involved in keratin filament aggregation *in vivo*. The dispersion step of filaggrin in the *fsn/fsn* epidermis appears to be defective, resulting in the accumulation of a large number of superficial cell layers with a filaggrin pattern characteristic of transition cells.[51,67]

Tape stripping of tail and dorsal skin revealed exaccerbation of the psoriasiform dermatitis on the dorsal skin but not tail skin of *fsn/fsn* mice. Control mouse skin from either site did not respond.[18] This so-called "Koebner" reaction is identified in 45% of human psoriatic patients.[68]

One of the values of mouse models is to test efficacy of novel compounds. To establish the usefulness of these models for this approach, the response in affected mice should mimic patients on more traditional treatments. Cyclosporin A is an immunosuppressive compound that is commonly used to treat severely psoriatic patients.[69] Flaky skin mice did not respond to either a subcutaneous cyclosporin A implant (5 mg/pellet, 45 day release pellets) or a placebo implant at either the gross or microscopic levels.[18] The same procedure was used on nude (*nu/nu*) mice, a mutation known to grow hair when treated with cyclosporin A.[70-72] There was no response in the nude mice. This study was repeated using topical applications of cyclosporin A on both *fsn/fsn* and *nu/nu* mice and their littermate controls. Again there was no response in any of the groups tested. We concluded that cyclosporin A was not effective in treating the epidermal hyperplasia in *fsn/fsn* mice by the routes or doses administered.[18] Steroids are well known to reduce inflammation. Intralesional injection of triamcinolone acetonide did not produce noticeable changes in the gross appearance of either mutant mice injected with the steroid or equal volumes of phosphate-buffered saline. Histologically, the dermal infiltrate was essentially eliminated (Sundberg, Boggess, and King, unpublished data).

Flaky skin mice, on the BALB/cByJ congenic background (N10+), are available from The Jackson Laboratory. Due to the anemia that the homozygotes develop, an average of 20% of the mutant mice die in shipment or shortly thereafter (King and Nanney, unpublished data).

E. FLAKY TAIL (GENE SYMBOL: *ft*)

The flaky tail mutation arose spontaneously at The Jackson Laboratory in 1958 among the progeny of crosses between heterogeneous stocks. Offspring developed abnormally small ears. When backcrossed onto the C57BL/6J strain, the offspring also developed tail constrictions and flaky tail skin, thus giving it the name flaky tail.[73] Flaky tail is an autosomal recessive mutation that maps to mouse Chromosome 3.[16]

Homozygous (*ft/ft*) mice can be recognized at 2 to 4 days of age by a stretched and shiny appearance of the skin covering the feet and back. At 5 to 14 days of age the tail develops annular constrictions with thick white flakes along its entire length (Figure 14). Shortened pinnae may be evident. After 14

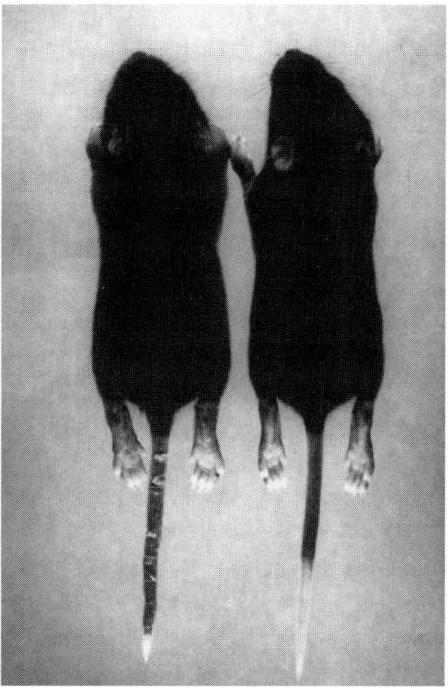

FIGURE 14   Flaky tail (*ft/ft*) mouse (left) and littermate control (+/*ft*, right) illustrate the marked scaling of the mutant mouse's tail skin at 2 weeks of age.

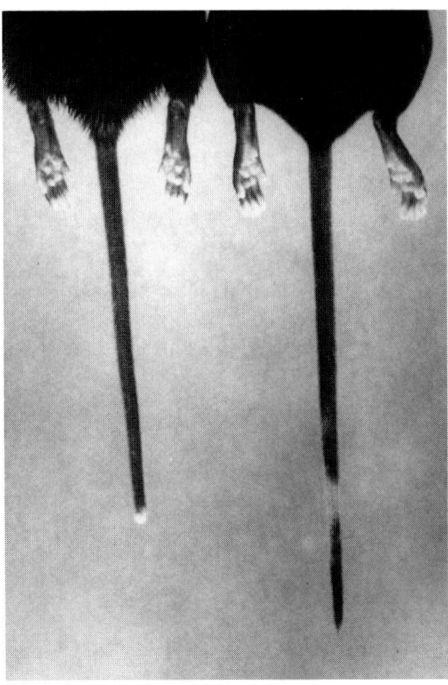

FIGURE 15   Autoamputation of the tail tip (*ft/ft*, left) and normal tail skin are features of the flaky tail mutation as the mice age (5 weeks of age). The rough hair coat (left) is the matted (*ma/ma*) phenotype, which is maintained on this background.

days of age, the tail flaking generally ceases, but the constrictions may result in autoamputation (ainhum; Figure 15). After 3 to 4 weeks of age mice appear to be normal except for the shortened ears and amputated tail tips.[73] Male and female flaky tail mice are fertile and live as long as controls. Some runted *ft/ft* mice die prior to weaning.[73]

The only histologic change reported in the skin of the ear, tail, and dorsal trunk in flaky tail mutant mice is a reduction of the thickness of the stratum granulosum in mutant mice compared to that of littermate control mice at ages 1, 2, 4, and 8 days. The author suggested that there was decreased keratinocyte proliferation in *ft/ft* mice, although no DNA synthesis studies were done.[73] Microscopically, the tail skin of 2-week-old *ft/ft* mice was found to be acanthotic and orthokeratic with large desquamating scales compared to the littermate controls (Figure 16).[74]

The flaky tail mutation is maintained at The Jackson Laboratory on a background that also has the matted (*ma*) mutation.[74] Phenotypic changes due to this latter mutation have to be differentiated from the former when studying flaky tail on the double mutant mouse. The STOCK *a/a ma ft/ma ft* is preserved in The Jackson Laboratory Frozen Embryo Repository. DNA extracts are available from the Mouse DNA Resource at The Jackson Laboratory.

FIGURE 16   The tail of *ft/ft* mice is mildly acanthotic and moderately orthokeratotic with marked desquamation at 2 weeks of age.

F.   ICHTHYOSIS (GENE SYMBOL: *ic*)

The ichthyosis mutation was identified in 1950 by Carter and Phillips.[75] Remutations at the ichthyosis locus have subsequently occurred, been identified, characterized, and are available on a variety of inbred backgrounds.[76,77] The ichthyosis mutation is autosomal recessive and maps to chromosome 1.[16]

The homozygous (*ic/ic*) mouse is smaller than normal inbred mice including heterozygous (+/*ic* or +/+) littermate controls. Mutant mice may be mute with retraction of the lower lip.[75] The pelage hair is sparse to almost nonexistent at the gross level (Figure 17). Vibrissae are short and coiled (Figure 18). Skin of the torso is often dry with a very fine white scale. Eyelids may be thickened and nails overgrown. Tails develop fine scales that become more prominent with age and develop annular constrictions resembling "ringtail." These features become exaggerated in a low humidity environment.[78,79] Severity of changes varies when the *ic* mutant gene is expressed on different genetic backgrounds.

Truncal epidermis is mildly thickened in *ic/ic* mice (Figure 19). Piliary canals are dilated and may contain bacterial colonies, particularly in older mice (Figure 20). The cuticle of the inner root sheath and hair shaft do not develop normally resulting in the formation of abnormal hair shafts within the piliary canal of both pelage hairs and vibrissae.[78,79]

Individual hairs are uneven in thickness. The medulla of the hair shaft is irregular with abnormal distribution of melanin granules. The regular dark/light banding pattern of the wild type mouse hair is absent, which can be exaggerated using polarized light.[80] Examination of the hair by scanning electron microscopy revealed defective or absent cuticles.[79] The cuticular defect corre-

FIGURE 17   An ichthyosis (*ic/ic*) mutant mouse (left) is smaller and has a thin hair coat compared with its normal littermate (+/*ic*, right).

FIGURE 18   Vibrissae are short and coiled on *ic/ic* mice (left) compared to the long straight vibrissae of a control (+/*ic*, right) mouse.

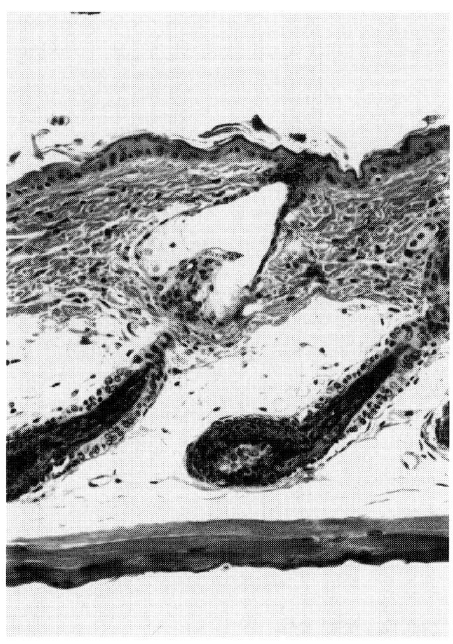

FIGURE 19   In adult *ic/ic* mice, hair follicles may be dilated.

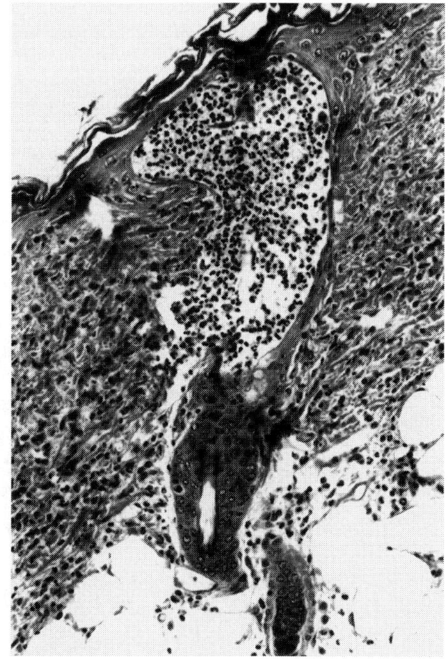

FIGURE 20   Older *ic/ic* mice develop suppurative inflammation with rupture of follicles.

sponds to a relative deficiency in the ultrahigh sulfur-containing amino acids (hard keratins) as determined by X-ray microanalysis and amino acid analysis.[81]

The Harderian gland, a specialized retroorbital exocrine gland found in rodents, may be absent.[82]

Dermal–epidermal recombination graft studies have demonstrated that the abnormal gene acts in the epidermis.[83] In addition, epidermis transplanted from *ic/ic* mice to normal mouse testicles revealed the development of abnormal hairs, indicating that the abnormal follicular development is due to abnormal gene function in the epidermis. The genotype of the dermis exerted no effect.[83]

Homozygous mutant mice have abnormally clumped chromatin in several cell types such as leukocytes and neurons.[84,85] Nuclei of cells from *ic/ic* mice have about 10% more DNA than do normal mice of other genotypes. Heterozygous (+/*ic*) mouse DNA content is between that of normal and *ic/ic* mice.[86]

Genital abnormalities, such as cryptorchidism in males and vaginal deformities in females, reduce the fecundity of ichthyosis mice.[80]

The ichthyosis mutation and remutations have been considered to be the "most useful of the animal models of ichthyosis for research purposes," although the authors went on to question its value as a model for any of the human ichthyoses.[87] A more recent review indicated that this mutation has several features of human ichthyosis vulgaris, lamellar ichthyosis, and psoriasis, but is not a model for any one of them.[79] The constrictions of the tail resemble the consequence of the inherited human disease ainhum or pseudoainhum, although the cause appears to be different.[79,80,88] Recent reevaluation of the morphological and biochemical features of the hair shafts indicates that the ichthyosis mouse mutation has many similarities to the inherited human hair diseases known as the trichothiodystrophies, a group of diseases that includes some patients with a type of ichthyosis.[89] The human disease is very pleomorphic[89,90] and more extensive biochemical analyses, which are in progress, are needed before the mouse mutation and human patients with specific types of trichothiodystrophy can be compared to determine which human subtype is the homologous disease.

The IC/Le inbred strain is maintained by The Jackson Laboratory Genetic Resource. A remutation of ichthyosis (*ic^J*) arose in the C57BL/6JEi strain in 1976.[91] Several additional remutations occurred at The Jackson Laboratory and are available on their strain of origin from the DNA Resource. The C57BL/6J *ic^J*-strain is preserved in The Jackson Laboratory Frozen Embryo Repository. DNA extracts are available from AKR/J-*ic* and IC/Le-*ic* strains from the Mouse DNA Resource at The Jackson Laboratory.

III. CONCLUSIONS

Many mouse mutations exist or can be manipulated to produce various forms of proliferative skin disease. Collectively these mutations provide a valuable resource to investigators interested in studying the pathophysiology of epidermal proliferation resulting in scaly skin types of diseases. Those mutations described here are or will be readily available to investigators and therefore they are the most valuable as models. Many new spontaneous and induced mutations are being characterized currently and offer a future filled with models that may cover most, if not all of the proliferative skin diseases known to affect humans and domestic mammals.

ACKNOWLEDGMENTS

This work was supported by grants from the National Alopecia Areata Foundation, National Institutes of Health (AR40324, RR8911, and CA34196, JPS; DK26518, LEK), and research funds from the Department of Veterans Affairs (LEK).

References

1. **Sundberg, J.P.,** Inherited mouse mutations: animal models and biomedical tools. *Lab. Anim.,* 20, 40, 1991.
2. **Sundberg, J.P.,** Conceptual evaluation of animal models as tools for the study of diseases in other species, *Lab. Anim.,* 21, 48, 1991.
3. **Sundberg, J.P.,** Inbred laboratory mice as animal models and biomedical tools: general concepts, in *Handbook of Mouse Mutations with Skin and Hair Abnormalities: Animal Models and Biomedical Tools*, Sundberg, J.P., Ed., CRC Press, Boca Raton, FL, 1994, 9.
4. **Sundberg, J.P. and Hogan, M.E.,** Hair types and subtypes in the laboratory mouse, in *Handbook of Mouse Mutations with Skin and Hair Abnormalities: Animal Models and Biomedical Tools*, Sundberg, J.P., Ed., CRC Press, Boca Raton, FL, 1994, 57.
5. **Chase, H.B. and Eaton, G.J.,** The growth of the hair follicles in waves, *Ann. N.Y. Acad. Sci.,* 83, 365, 1959.
6. **Straile, W.E., Chase, H.B., and Arsenault, C.,** Growth and differentiation of hair follicles between periods of activity dend quiescence, *J. Exp. Zool.,* 148, 205, 1961.
7. **Chase, H.B.,** The physiology and histochemistry of hair growth, *J. Soc. Cosmetic Chem.,* 6, 9, 1955.
8. **Sundberg, J.P., Hogan, M.E., and King, L.E., Jr.,** Normal biology and aging changes in mouse skin and hair, in *Pathobiology of the Aging Mouse*, Mohr, U., Ed., ILSI Press, Washington, D.C., in press.

9. **Silvers, W.K.,** *The Coat Colors of Mice. A Model for Mammalian Gene Action and Interaction,* Springer-Verlag, New York, 1979.

10. **Orlow, S.J.,** Aging of the murine pigmentary system in normal and pathologic states, in *Pathobiology of the Aging Mouse,* Mohr, U., Ed., ILSI Press, Washington, D.C., in press.

11. **Sundberg, J.P., Orlow, S.J., Sweet H.O., and Beamer, W.G.,** The adrenocortical dysplasia (*acd*) mutation, Chromosome 8, in *Handbook of Mouse Mutations with Skin and Hair Abnormalities: Animal Models and Biomedical Tools,* Sundberg, J.P., Ed., CRC Press, Boca Raton, FL, 1994, 159.

12. **Dunn, T.B.,** Morphology and histogenesis of mammary tumors, in *A Symposium on Mammary Tumors in Mice,* Moulton, F.R., Ed., American Association for the Advancement of Science, Washington, D.C., 1945.

13. **Sundberg, J.P., Boggess, D., Sundberg, B.A., Beamer, W.G., and Shultz, L.D.,** Epidermal dendritic cell populations in the flaky skin mutant mouse, *Immunol. Invest.,* 22, 389, 1993.

14. **Sundberg, J.P. and Orlow, S.J.,** The depilated (*dep*) mutation, chromosome 4, in *Handbook of Mouse Mutations with Skin and Hair Abnormalities: Animal Models and Biomedical Tools,* Sundberg, J.P., Ed., CRC Press, Boca Raton, FL, 1994, 235.

15. **Sundberg, J.P. and King, L.E., Jr.,** Mouse mutations as animal models and biomedical tools for dermatological research, *J. Invest. Dermatol.,* in press.

16. **Green, M.,** Catalog of mutant genes and polymorphic loci, in *Genetic Variants and Strains of the Laboratory Mouse,* 2nd ed., Lyon, M.F. and Searle, A.G., Eds., Oxford University Press, Oxford, 1989, 12.

17. **Sundberg, J.P.,** Mouse models for genetically based skin and hair diseases, *Dermatol. News, Jax Notes Suppl.,* 1993.

18. **Sundberg, J.P., Boggess, D., Shultz, L.D., and Beamer, W.G.,** The flaky skin (*fsn*) mutation, Chromosome?, in *Handbook of Mouse Mutations with Skin and Hair Abnormalities: Animal Models and Biomedical Tools,* Sundberg, J.P., Ed., CRC Press, Boca Raton, FL, 1994, 253.

19. **Blessing, M., Nanney, L.B., King, L.E., Jr., and Hogan, B.L.M.,** Chemical skin carcinogenesis is prevented in mice by induced expression of a TGF-B related transgene, *Teratogen. Carcinogen. Mutagen.,* in press.

20. **Sundberg, J.P., Brown, K.S., and McMahon, W.M.,** Chronic ulcerative dermatitis in black mice, in *Handbook of Mouse Mutations with Skin and Hair Abnormalities: Animal Models and Biomedical Tools,* Sundberg, J.P., Ed., CRC Press, Boca Raton, FL, 1994, 485.

21. **Sundberg, J.P., Sundberg, B.A., and King, L.E., Jr.,** Cutaneous aging changes in commonly used inbred mouse strains and mutant stocks, in *Pathobiology of Aging Mice,* Mohr, U., Ed., ILSI Press, Washington, D.C., in press.

22. **Gates, A.H. and Karasek, M.,** Hereditary absence of sebaceous glands in the mouse, *Science,* 148, 1471, 1965.

23. **Sweet, H.O. and Lane, P.W.,** Asebia-J on chromosome 19, *Mouse News Lett.,* 57, 20, 1977.

24. **Festing, M.F.W.,** Origin and characteristics of inbred strain s of mice. 14th listing, *Mouse Genome,* 90, 231, 1992.

25. **Compton, J.G., Dunstan, R.W., and Sundberg, J.P.,** Asebia, a mutation affecting sebaceous glands, mouse, in *Monographs on Pathology of Laboratory Animals. Integument and Mammary Glands,* Jones, T.C., Mohr, U., and Hunt, R.D., Eds., Springer-Verlag, Heidelberg, 1989, 218.

26. **Pennycuik, P.R., Raphael, K.A., Chapman, R.E., and Hardy, M.H.,** The site of action of the asebia locus (*ab*) in the skin of the mouse, *Genet. Res.*, 48, 179, 1986.

27. **Josefowicz, W.J. and Hardy, M.H.,** The expression of the gene asebia in the laboratory mouse. 3. Sebacous glands, *Genet. Res.*, 31, 157, 1978.

28. **Sundberg, J.P.,** The asebia (*ab, ab^J*) mutations, Chromosome 19, in *Handbook of Mouse Mutations with Skin and Hair Abnormalities: Animal Models and Biomedical Tools,* Sundberg, J.P., Ed., CRC Press, Boca Raton, FL, 1994, 171.

29. **Josefowicz, W.J. and Hardy, M.H.,** The expression of the gene asebia in the laboratory mouse. I. Epidermis, dermis, *Genet. Res.*, 31, 53, 1978.

30. **Brown, W.R. and Hardy, M.H.,** A hypothesis on the cause of chronic epidermal hyperproliferation in asebia mice, *Clin. Exp. Dermatol.*, 13, 74, 1988.

31. **Josefowicz, W.J. and Hardy, M.H.,** The expression of the gene asebia in the laboratory mouse. II. Hair follicles, *Genet. Res.*, 31, 145, 1978.

32. **Brown, W.R. and Hardy, M.H.,** Mast cells in asebia mouse skin, *J. Invest. Dermatol.*, 93, 708, 1989.

33. **Wilkinson, D.I. and Karasek, M.A.,** Skin lipids of a normal and a mutant (asebic) mouse strain, *J. Invest. Dermatol.*, 47, 449, 1966.

34. **Brown, W.R., Furukawa, R.D., and Ramsay, C.A.,** Circadian rhythms are suppressed in hyperproliferative mouse epidermis, *Cell Tissue Kinet.*, 21, 159, 1988.

35. **Brown, W.R., Furukawa, R.D., and Ramsay, C.A.,** UVB radiation further stimulates proliferation in the already hyperproliferative epidermis of the asebia mouse, *Arch. Dermatol. Res.*, 281, 366, 1989.

36. **Sundberg, J.P., Beamer, W.G., Shultz, L.D., and Dunstan, R.W.,** Inherited mouse mutations as models of human adnexal, cornification, and papulosquamous dermatoses, *J. Invest. Dermatol.*, 95, 62S, 1990.

37. **Lyons, M.F., Phillips, R.J.S., and Bailey, H.J.,** Mutagenic effects of repeated small radiation doses to mouse spermatogonia. I. Specific locus mutation rates, *Mut. Res.*, 15, 185, 1972.

38. **Phillips, R.J.S., Hawker, S.G., and Moseley, H.J.,** Bare-patches, a new sex-linked gene in the mouse, associated with a high production of XO females. I. A preliminary report of breeding experiments, *Genet. Res.*, 22, 91, 1973.

39. **Happle, R., Phillips, R.J.S., Roessner, A., and Junemann, G.,** Homologous genes for X-linked chondrodysplasia punctata in man and mouse, *Hum. Genet.*, 63, 24, 1983.

40. **Falconer, D.S. and Isaacson, J.H.,** Harlequin and brindled-linkage, and difference between coupling and repulsion phenotypes, *Mouse News Lett.*, 47, 28, 1972.

41. **Herman, G.E. and Walton, S.J.,** Close linkage of the murine locus bare patches to the X-linked visual pigment gene: implications for mapping human X-linked dominant chondrodysplasia punctata, *Genomics*, 7, 307, 1990.

42. **Chatterjee, A., Faust, C.J., and Herman, G.E.,** Genetic and physical mapping of the biglycan gene on the mouse X chromosome, *Mamm. Genome*, 4, 33, 1993.

43. **Phillips, R.J.S. and Kaufman, M.H.,** Bare-patches, a new sex-linked gene in the mouse, associated with a high production of XO females. II. Investigations into the nature and mechanism of the XO production, *Genet. Res.*, 24, 27, 1974.

44. **Phillips, R.J.S.,** Factor for XO production — In(X)1H, *Mouse News Lett.*, 52, 35, 1975.

45. **Evans, E.P. and Phillips, R.J.S.,** Inversion heterozygosity and the origin of XO daughters in *Bpa/+* female mice, *Nature (London)*, 256, 40, 1975.

46. **Happle, R.,** X-linked dominant chondrodysplasia punctata. Review of the literature and report of a case, *Hum. Genet.*, 53, 65, 1979.

47. **Happle, R.,** Cataracts as a marker of genetic heterogeneity in chondrodysplasia punctata, *Clin. Genet.*, 19, 64, 1981.

48. **Traupe, H., Muller, D., Atherton, D., Kalter, D.C., Cremers, F.P.M., van Oost, B.A., and Ropers, H.-H.,** Exclusion mapping of the X-linked dominant chondrodysplasia punctata/ichthyosis/cataract/short stature (Happle) syndrome: possible involvement of an unstable pre-mutation, *Hum. Genet.*, 89, 659, 1992.

49. **HogenEsch, H., Gijbels, M.J.J., Offerman, E., van Hooft, J., van Bekkum, D.W., and Zurcher, C.,** A spontaneous mutation characterized by chronic proliferative dermatitis in C57BL mice, *Am. J. Pathol.*, 143, 972, 1993.

50. **Gijbels, M.J.J., HogenEsch, N., Blauw, B., Roholl, P., and Zurcher, C.,** Ultrastructure of epidermis of mice with chronic proliferative dermatitis, *Ultrastr. Pathol.*, 19, 107, 1995.

51. **Morita, K., Hogan, M.E., Nanney, L.B., King, L.E., Monabe, M., Sun, T.-T., and Sundberg, J.P.,** Cutaneous ultrastructural features of the flaky skin (*fsn*) mouse mutation, *J. Dermatol.*, 22, 8, 1995.

52. **Raick, A.N.,** Ultrastructural, histological, and biochemical alterations produced by 12-O-tetradecanoyl-phorbol-13-acetate on mouse epidermis and their relevance to skin tumor promotion, *Cancer Res.*, 33, 269, 1973.

53. **Feldman, D., Bryce, G.F., and Shapiro, S.S.,** Mitochondrial inclusions in keratinocytes of hairless mouse skin exposed to UVB radiation, *J. Cutan. Pathol.*, 17, 96, 1990.

54. **Gijbels, M.J.J., HogenEsch, N., Bruijnzeel, P.L.B., Elliott, G.R., and Zurcher, C.,** Maintenance of donor phenotype after full-thickness skin transplantation from mice with chronic proliferative dermatitis (*cpdm/cpdm*) to C57BL/Ka and nude mice and vice versa, *J. Invest. Dermatol.*, in press.

55. **Beamer, W.G., Maltais, L.J., and Bernstein, S.E.,** Disturbed iron metabolism in flaky skin (*fsn/fsn*) mutant mice, *Jackson Lab. Ann. Sci. Rep.*, 92, 1986.

56. **Beamer, W.G., Shultz, L.D., Barker, J.E., and Sundberg, J.P.,** Iron deficiency anemia in flaky skin (*fsn/fsn*) mutant mice, *Blood*, in press.

57. **Sundberg, J.P., Boggess, D., Beamer, W.G., and Shultz, L.D.,** Role of hematopoietic progenitor cells in the development of psoriasiform dermatitis in the flaky skin mutant mouse, *J. Invest. Dermatol.*, 98, 558, 1992.

58. **Compton, J.G., Dunstan, R.W., Beamer, W.G., Roop, D.R., and Sundberg, J.P.,** Flaky skin: a new mouse mutation developing a heritable papulosquamous disease, *J. Invest. Dermatol.*, 92, 414, 1989.

59. **Russel, E.S.,** Hereditary anemias of the mouse: a review for geneticists, *Adv. Genet.*, 20, 357, 1979.

60. **Jahn, H., Nielsen, E.H., Elberg, J.J., Bierring, R., Ronne, M., and Brandrup, F.,** Ultrastructure of psoriatic epidermis, *APMIS*, 96, 723, 1988.

61. **Heng, M.C.Y., Heng, J.A., and Allen, S.G.,** Electron microscopic features in generalized pustular psoriasis, *J. Invest. Dermatol.*, 89, 187, 1987.

62. **Heng, M.C.Y., Kloss, S.G., Kuehn, C.S., and Chase, D.G.,** Significance and pathogenesis of basal keratinocyte herniations in psoriasis, *J. Invest. Dermatol.,* 87, 362, 1986.

63. **Frei, J.V. and Sheldon, H.,** Corpus intra cristam: a dense body within mitochondria of cells in hyperplastic mouse epidermis. *J. Biophys. Biochem. Cytol.,* 11, 721, 1961.

64. **Sundberg, J.P., Kenty, G.A., Beamer, W.G., and Adkison, D.L.,** Forestomach papillomas in flaky skin and steele-Dickie mutant mice, *J. Vet. Diag. Invest.,* 4, 312, 1992.

65. **Sundberg, J.P. and Shultz, L.D.,** Severe combined immunodeficiency, *J. Comp. Pathol.,* 25, 3, 1993.

66. **Sundberg, J.P., Dunstan, R.W., Roop, D.R., and Beamer, W.G.,** Full-thickness skin grafts from flaky skin mice to nude mice: maintenance of the psoriasiform phenotype, *J. Invest. Dermatol.,* 102, 781, 1994.

67. **Manabe, M., Dale, B., Sundberg, J.P., and Sun, T.-T.,** Maturation pathways of filaggrin and abnormal epidermis, *J. Invest. Dermatol.,* 94, 552, 1990.

68. **Krueger, G.G.,** A perspective on psoriasis as an aberration in skin modified to expression by the inflammatory/repair system, in *Immune Mechanisms in Cutaneous Disease,* Norris, D.A., Ed., Marcel Dekker, New York, 1989, 425.

69. **Mihatsch, M.J. and Wolff, F.F.K.,** Risk/benefit ratio of cyclosporin A (Sandimmune) in psoriasis, *Br. J. Dermatol.,* 122(Suppl. 36), 1, 1990.

70. **Pendry, A. and Alexander, P.,** Stimulation of hair growth on nude mice by cyclosporin A, in *Cyclosporin A,* White, J.G., Ed., Elsevier, Amsterdam, 1982, 77.

71. **Sawada, M., Terada, N., Taniguchi, H., Tateishi, R., and Mori, Y.,** Cyclosporin A stimulates hair growth in nude mice, *Lab. Invest.,* 56, 684, 1987.

72. **Buhl, A.E., Waldon, D.J., Miller, B.F., and Brunden, M.N.,** Differences in activity of minoxidil and cyclosporin A on hair growth in nude and normal mice, *Lab. Invest.,* 62, 104, 1990.

73. **Lane, P.W.,** Two new mutations in linkage group XVI of the house mouse, *J. Hered.,* 63, 135, 1972.

74. **Sundberg, J.P.,** The flaky tail (*ft*) mutation, Chromosome 3, in *Handbook of Mouse Mutations with Skin and Hair Abnormalities: Animal Models and Biomedical Tools,* Sundberg, J.P., Ed., CRC Press, Boca Raton, FL, 1994, 269.

75. **Carter, T.C. and Phillips, R.S.,** Ichthyosis, a new recessive mutant in the house mouse, *J. Hered.,* 41, 297, 1950.

76. **Lane, P.W., Mobraaten, L.E., and Neleski, L.A.,** *Lists of Mutations and Mutant Stocks of the Mouse Maintained at The Jackson Laboratory,* The Jackson Laboratory, Bar Harbor, ME, 1992.

77. **Sundberg, J.P. and Shultz, L.D.,** Inherited mouse mutations: models for the study of alopecia, *J. Invest. Dermatol.,* 96, 95s, 1991.

78. **Sundberg, J.P. and Pittelkow, M.R.,** The ichthyosis (*ic*) mutation, Chromosome 1, in *Handbook of Mouse Mutations with Skin and Hair Abnormalities: Animal Models and Biomedical Tools,* Sundberg, J.P., Ed., CRC Press, Boca Raton, FL, 1994, 327.

79. **Holbrook, K.A.,** Ichthyosis, inherited, skin, mouse (ic/ic), in *Monographs on the Pathology of Laboratory Animals. Integument and Mammary Glands,* Jones, T.C., Mohr, U., and Hunt, R.D., Eds., Springer-Verlag, Heidelberg, 1989, 223.

80. **Spearman, R.I.,** The skin abnormality of "ichthyosis", a mutant of the house mouse, *J. Embryol. Exp. Morphol.*, 8, 387, 1960.
81. **Itin, P.H., Sundberg, J.P., Dunstan, R.W., and Pittelkow, M.R.,** Ichthyosis mouse-structural and biochemical analysis of the hair defect, *J. Invest. Dermatol.*, 94, 537, 1990.
82. **Gruneberg, H.,** Apocrine glands and the Chizvitz organ of some mouse mutants, *J. Embryol. Exp. Morphol.*, 25, 247, 1971.
83. **Green, M.C., Alpert, B.N., and Mayer, T.C.,** The site of action of the ichthyosis locus (*ic*) in the mouse, as determined by dermal-epidermal recombinations, *J. Embryol. Exp. Morphol.*, 32, 715, 1974.
84. **Green, M.C., Shultz, L.D., and Nedzi, L.A.,** Abnormal nuclear morphology of leukocytes in the mouse mutant ichthyosis. A possible transplantation marker, *Transplantation*, 20, 172, 1975.
85. **Goldowitz, D. and Mullen, R.J.,** Nuclear morphology of ichthyosis mutant mice as a cell marker in chimeric brain, *Dev. Biol.*, 89, 261, 1982.
86. **Meyers, R.S., Klein, A.S., Eppig, J.J., and Eckhardt, R.A.,** Altered DNA content and chromatin distribution associated with the mouse ichthyosis gene, *Genetics*, 83, s50, 1976.
87. **Holbrook, K.A. and Sybert, V.P.,** Animal models of inherited ichthyoses and other forms of aberrant epidermal differentiation, in *Models in Dermatology*, Vol. I, Maibach, H.I., and Lowe, N., Eds., Karger AG, Basel, 1985, 132.
88. **Selmanowitz, V.J.,** Ectodermal dysplasias including epitheliogenesis imperfecta, ichthyosis, and follicle/glandular anomalies, in *Spontaneous Models of Human Diseases*, Vol. II, Andrews, E.J., Ward, B.C., and Altman, N.H., Eds., Academic Press, New York, 1979, 3.
89. **Itin, P.H. and Pittelkow, M.R.,** Trichothyodystrophy: review of sulfur-deficient brittle hair syndromes and association with the ectodermal dysplasias, *J. Am. Acad. Dermatol.*, 22, 705, 1990.
90. **Itin, P.H. and Pittelkow, M.R.,** Trichothiodystrophy with chronic neutropenia and mild mental retardation, *J. Am. Acad. Dermatol.*, 24, 356, 1991.
91. **Eicher, E.,** Remutations, *Mouse News Lett.*, 54, 40, 1976.

7 Whole Skin Organ Culture as a Tool to Study Hyaluronan and Proteoglycan Metabolism

*Raija Tammi, Anna-Liisa Tuhkanen-Martikainen, Ulla Ågren,
and Markku Tammi*

TABLE OF CONTENTS

I. INTRODUCTION

Proteoglycans (PGs)[*] are a functionally diverse group of macromolecules that consists of a protein core to which one or more glycosaminoglycan (GAG) chain(s) are attached (for reviews, see References 1 to 3). PGs are found in extracellular matrices, on cell surfaces, and in intracellular granules. Cell-surface-associated PGs usually contain heparan sulfate (HS) and/or chondroitin sulfate (CS). Matrix associated PGs contain mostly DS, CS, and/or KS chains. Hyaluronan (HA) is an unsulfated GAG usually not covalently bound to a protein core, but it shows high affinity binding to certain proteins like HA binding PGs, fibronectin, and Type VI collagen.[4] The importance of PGs and HA in the maintenance of the hydrated and elastic extracellular matrix of cartilage and skin has been long acknowledged,[5] but it is becoming more and more apparent that these molecules also have more specific functions in the regulation of cell proliferation and differentiation (see reviews, see References 2 to 4).

Adult human skin is one of the richest sources of GAGs in the body.[6-8] HA and DS form 50 and 40%, respectively, of GAGs in skin, while CS and HS constitute less than 10%.[7,8] Although the vast majority of skin GAGs are found in the dermis, the epithelial compartment has its distinct content and distribution of GAGs.[9-12] Biochemical characterizations of PGs have revealed that decorin (a small PG carrying one DS chain) is the major PG present in skin,[13-18] while smaller amounts of another small DS-PG (biglycan)[19] and two large DS/CS PGs[20,21] are also present. Immunohistochemical stainings have shown that decorin is localized throughout the dermal tissue, with the highest concentration found in papillary dermis and around skin appendages,[22-25] while biglycan[26] and PG-100 (a third small DS-PG)[27] were found mainly in epidermis. Versican, a large HA-binding CS-PG,[28] is localized in adult skin in the basal cell layer of the epidermis and around elastic fibers in the dermis, while its distribution is more widespread in embryonic dermis.[29] Immunohistochemical and *in situ* hybrid-

---

[*] Abbreviations: BSA, bovine serum albumin; CHAPS 3-[(3-cholamidopropyl)dimethylammonio]-1-propanesulfonate; CD44, phagocytic glycoprotein-1 (Pgp-1), Hermes-antigen, p85 protein, extracellular matrix receptor type III, hyaluronate receptor; CPC, cetylpyridinium chloride; CS, chondroitin sulfate; CS-PG, chondroitin sulfate proteoglycan; DS, dermatan sulfate; DS-PG, dermatan sulfate proteoglycan; GAG, glycosaminoglycan; HA, hyaluronan, hyaluronate, hyaluronic acid; HABC, hyaluronan binding region, and link protein complex; bHABC, biotinylated hyaluronan binding region, and link protein complex; HABR, hyaluronan binding region (G$_1$) of aggrecan; HS, heparan sulfate; HS-PG, heparan sulfate proteoglycan; KS, keratan sulfate; KS-PG, keratan sulfate proteoglycan; NEM, *N*-ethylmaleimide; PBS, phosphate-buffered saline; PG, proteoglycan; PMSF, phenylmethylsulfonyl fluoride; sGAG, sulfated glycosaminoglycan.

ization techniques have verified that epidermal and dermal cells express various HS-PGs (members of the syndecan family, glypican, CD44, and perlecan),[12,30-34] but little is known of the chemical quantity or tissue metabolism of these molecules.

The expression of the PGs and HA has been extensively studied using cultures of skin fibroblasts[25,35-41] and epidermal keratinocytes.[42-47] Cell culture experiments have been most helpful in detecting and characterizing the repertoire of PGs the cells are to synthesize. However, in culture, cells are more or less dedifferentiated, perhaps due to a loss of normal cell–matrix and cell–cell contacts. The cell proliferation rate and expression of differentiation markers do not generally correspond to those under *in vivo* conditions. This may lead to the altered profile of PGs expressed in cell culture compared to the cells *in situ*. Not only the synthesis but also the degradation of PGs may be markedly modulated by the lack of a normal matrix.[48,49]

Organ culture offers an alternative to cell culture models. In organ culture, epidermal cells continue to proliferate, stratify, and differentiate in an ordered fashion, while the dermal cells remain embedded in their normal three-dimensional matrix, and exhibit low levels of proliferation while continuing to be metabolically highly active.[50-52] Short-term organ cultures have been utilized to study HA and PG metabolism in pig and rat skin.[21,53] Here we describe examples of experiments on the synthesis and metabolism of HA and PGs in adult human skin organ culture, using metabolic labeling of GAGs with [³H]glucosamine and $^{35}SO_4$ and analyzing the GAGs and PGs after separating epidermis and dermis.

II. MATERIALS AND METHODS

A. ORGAN CULTURE

Adult human skin was cultured under chemically defined conditions.[50] Skin was dissected into small (2 to 3 mm²) pieces that were put on grids covered with lens paper in contact with medium in small petri dishes. Epidermis of the explants was exposed in the air. Eagle's minimum essential medium (MEM), supplemented with L-glutamine (2mM) and antibiotics (penicillin 50 U/ml and streptomycin 50 µg/ml) without serum was used as a culture medium. The petri dishes were incubated in airtight, humified chambers aerated daily with a special gas mixture (40% O_2, 5% CO_2, and 55% N_2) at +37°C and 100 µCi/ml $^{35}SO_4$ or 5 µCi/ml of [³H]glucosamine (Amersham, U.K.) was added to culture medium to label the GAGs synthesized during culture for 16 to 24 h. After labeling epidermis and dermis were separated by a brief treatment with 0.04% EDTA at 37°C.[54]

For pulse-chase experiments, explants were incubated with 5 µCi/ml of [³H]glucosamine for 24 h on the first culture day, washed with unlabeled culture medium, and transferred to organ culture. After 24 h chase, the cultures were terminated, and tissues processed as described above.

B. EXTRACTION OF [³H]GLUCOSAMINE-LABELED GAGS

Epidermis and dermis were dried in acetone at 4°C, and GAGs were liberated with papain (250 µg/ml) in 5 mM cysteine and 5 mM EDTA at 60°C for 24 h.[54] After digestion, the samples were boiled for 15 min to inactivate the enzyme, and the digestion mixtures were cleared either by centrifugation or by filtration through 0.8-µm nitrocellulose membranes.

C. EXTRACTION OF PROTEOGLYCANS

The PGs labeled with [³H]glucosamine or $^{35}SO_4$ were extracted with 4M guanidinium chloride and 0.1% CHAPS in 12 mM sodium acetate buffer, pH 5.8 in the presence of protease inhibitors (6 mM EDTA, 25 mM 6-aminohexanoic acid, 250 mM benzamidine-HCl, 2.5 mM NEM, and 1 mM PMSF) for 24 h at +4°C.[55] The samples were centrifuged and the supernatants saved. The tissues (extraction residues) were washed with an equal volume of the extraction buffer, and the wash was combined with the extract.

D. QUANTITATION OF [³H]GLUCOSAMINE INCORPORATION INTO HA AND PGS

The incorporation of activity in GAGs was assayed by using either a method published by Saami and Tammi[56,57] or that of Ågren et al.[56] In the former method, GAGs were precipitated onto a nitrocellulose membrane with 1% cetylpyridinium chloride (CPC),[57] and HA was specifically eluted with 0.5 M HCl to separate it from sGAGs that remain on the membrane. The radioactivities in the eluents and membranes were separately counted. In the latter method predigestion with 24 TRU/ml of *Streptomyces* hyaluronidase (Seikakagu Kogyo CO., Japan) was used to differentiate between hyaluronan and sulfated GAGs. Parallel aliquots of the samples, with and without digestion, were mixed with CPC (final concentration 1%) and pipetted into the wells of a dot-blot apparatus. The precipitates formed were collected onto nitrocellulose membranes by vacuum. Unincorporated precursor and glycopeptides were washed through with distilled water. The membranes were air-dried, and the individual dots cut for scintillation counting.

E. PURIFICATION OF HA AND PGS WITH ANION-EXCHANGE CHROMATOGRAPHY

Guanidinium chloride extracts of [³H]glucosamine-labeled tissues were purified using anion-exchange chromatography as described earlier.[58] The extracts were passed through a column of Sephadex G-25 (Pharmacia, Sweden) to remove unbound precursors and to change the buffer into 50 mM Tris-HCl, pH 8.0 with 0.5% CHAPS for subsequent purification on ion-exchange chromatography. Of each extract 900 µl was injected and pumped at 0.5 ml/min into a MA7P+ column (Bio-Rad, Hercules, CA). The column was eluted with a linear gradient of 0 to 3 *M* NaCl in the same buffer, and the eluent collected in 250-µl fractions. The NaCl gradient was monitored by conductivity, and the elution positions of standard HA (Healon, Pharmacia) and chondroitin sulfate (Sigma, St. Louis, MO) were determined by the Safranin O assay.[59]

The identities of HA and chondroitin sulfate were confirmed by digesting with *Streptomyces* hyaluronidase (24 U/ml in 0.15 *M* Na-acetate buffer, pH 5.0 for 20 h at +37°C) and chondroitinase ABC (50 mU in 0.1 *M* Tris-HCl buffer, pH 7.3 for 6 h at +37°C), respectively. The identity of heparan sulfate was confirmed using degradation with 0.24 *M* nitrous acid in 1.8 *M* acetic acid for 80 min at room temperature.[60] The samples were examined before and after the chemical or enzymic degradation by comparing their elution patterns on gel filtration (Sephadex G-50 Fine, Pharmacia).

F. CHROMATOGRAPHY OF HA ON SEPHACRYL S-1000

The fractions of MAP7+ anion-exchange chromatography that contained HA were pooled and used for molecular mass determination on a 1 × 30-cm column of Sephacryl S-1000 (Pharmacia) eluted with a sodium acetate buffer (0.15 *M*, pH 6.8) containing 0.5% CHAPS. Fractions (0.5 ml) were collected and counted for radioactivity.[58]

G. SDS-AGAROSE GEL ELECTROPHORESIS OF $^{35}SO_4$-LABELED PGs

Tissue extracts were dialyzed against 4 *M* urea and 0.1% SDS, in 40 m*M* Tris-acetate buffer, pH 6.8, followed by precipitation of PGs with 85% (v/v) ethanol containing 50 m*M* Na-acetate overnight at -80°C.

The precipitates were dissolved in a buffer containing 5 m*M* Ca-acetate (pH 7.0), 20% ethanol, and enzyme inhibitors (10 m*M* NEM, 1 m*M* PMSF, 0.5 m*M* Pepstatin A). The solutions were subjected to enzyme degradation of the PGs by treatment with heparitinase (70 mU/ml), chondroitinase ABC (140 mU/ml), keratanase (140 mU/ml), and a mixture containing both heparitinase and chondroitinase ABC (70 and 140 mU/ml, respectively).[55] All enzymes were from Seikagaku Kogyo Co (Japan). Controls were incubated under the same conditions but without any enzyme addition. After incubation, the samples were concentrated by precipitation in 85% (v/v) ethanol, dissolved in 40 m*M* Tris-acetate buffer, pH 6.8, containing 2% sodium dodecyl sulfate (SDS) and 1 m*M* Na_2SO_4, and boiled for 5 min. An equal volume of 60% sucrose (w/v) with 0.05% bromphenol blue was added before electrophoresis.

Commercial ^{14}C-labeled protein standards (Amersham), chondroitin sulfate C (Sigma), and large aggregating PG from bovine articular cartilage (A1-PG, a gift from Dr. A. M. Säämänen) were used as molecular mass standards. The samples were analyzed in submerged 0.7% (w/v) agarose slab gels by electrophoresis as described earlier.[55] Gels 3 mm thick were prepared in the electrophoresis buffer, which contained 0.03% SDS, 40 mM Tris-acetate, and 1 mM Na_2SO_4. Electrophoresis was done at room temperature for 3 h, using a constant current of 50 mA (38 to 40 V). The gels were fixed in methanol/acetic acid/water (50:7:43, v/v/v) for 1.5 h, dried, and then apposed onto autoradiographic film (Hyperfilm β Max, Amersham, U.K.).

H. QUANTITATION OF HYALURONAN

HA content was measured in triplicate with an "ELISA-like" method using biotinylated hyaluronan-binding probe (bHABC) prepared from the PG-link protein hyaluronan complex of bovine articular cartilage.[12] The assay method used was slightly modified from that of Kongtawelert and Ghosh.[61,62] Microwell plates were precoated with 50 μg/ml poly-L-lysine and air-dried. HA (100 μl, 50 μg/ml) was added to the plates and incubated for 1 h at 37°C. The plates were washed 3×5 min with phosphate-buffered saline (PBS) containing 0.05% Tween 20 (Tween-PBS) and blocked with 1% bovine serum albumin for 1 h at 37°C.

Standard hyaluronan (Healon, Pharmacia) containing 5 to 200 ng/ml of HA and samples were diluted into 6% BSA in PBS and mixed with an equal volume of bHABC. Samples and standards were incubated overnight at 4°C, then 1 h at 37°C, and 100 μl of each was transferred to HA-precoated wells, and incubated for 1.5 h at 37°C. After washes with PBS, 100 μl of avidin–biotin–peroxidase complex (Vector, CA) was added and allowed to stand for 1 h at room temperature. After washing 5×5 min with Tween-PBS the plates were air-dried. Substrate-chromogen solution (100 μl, 1 mg/ml of O-phenylenediamine dihydrochloride, 0.03% H_2O_2 in 0.1 M citrate-phosphate buffer, pH 5) was added and kept for 1 h at 37°C. The reaction was stopped by adding 50 μl of 8 M H_2SO_4, and the absorbances were read at 490 nm using a microtiter-plate reader. Each sample and standard was done in triplicate, and the mean of these values was used for the calculation of the final results.

III. RESULTS

A. DISTRIBUTION OF [3H]GLUCOSAMINE-LABELED GAGS IN SKIN ORGAN CULTURE

Table 1 shows the amount of [3H]glucosamine incorporated into GAGs in whole skin organ culture labeled on the fifth culture day. Both epidermis and dermis actively incorporate the precursor into HA and sGAGs. Based on dry

TABLE 1

Incorporation of [³H]Glucosamine into GAGs in Human Skin Organ Culture[a]

	HA		sGAGs	
	cpm/mg	%	cpm/mg	%
Epidermis	8500 ± 1100	62	5300 ± 680	38
Dermis	5400 ± 1100	58	3900 ± 690	42
Medium	2920 ± 1090[b]	76	900 ± 240[b]	24

[a] The tissue explants were labeled on the fifth culture day with [³H]glucosamine. Epidermis and dermis were separated, dried with acetone, and weighed. The GAGs were liberated with papain digestion, precipitated onto nitrocellulose membrane with CPC, and HA was specifically eluted with HCl. The activities recovered in eluent (HA) and that remaining on the membrane (sGAGs) were counted by liquid scintillation and calculated per tissue dry weight. The data represent the means ± SEM from eight independent cultures.

[b] Expressed per dermal dry weight.

weight, the rate of synthesis is higher in epidermis than in dermal tissue. However, because of the abundance of dermal tissue compared to epidermis, the former is responsible for the majority of total GAG production in organ culture (Table 1). In both epidermis and dermis, sulfated GAGs formed about 40% of total [³H]glucosamine-labeled GAGs in a 24-h culture, the remaining being HA (Table 1). During labeling, about 35% of the total [³H]HA but less than 20% of [³H]sGAGs were liberated into the culture medium. Because the explants are cultured above the level of the medium and diffusion from epidermis is limited by the basal lamina, the GAGs in the medium probably originate from the dermal compartment. Thus, relatively more of the newly synthesized HA diffuses into medium as compared to PGs.

B.   GAG TURNOVER IN SKIN ORGAN CULTURE

The disappearance rates of GAGs from epidermis and dermis were studied using a 24 h pulse of [³H]glucosamine on the first culture day and counting the remaining activity in the tissues after a 24-h chase. [³H]HA rapidly disappeared from both epidermal and dermal compartments (Figure 1). The estimated half-life of HA was about 1 day in both tissues. HA accumulated in the medium during the chase, and the amount in this compartment roughly corresponded to the amount lost from the dermal tissue during the chase (Figure 1).

The amount of the HA in cultured skin was measured using an ELISA-like binding assay with bHABC. After 5 days in culture epidermal HA content was slightly lower than the *in vivo* value,[63] while the dermal HA concentration at this time was considerably lower than the *in vivo* value[58] (Table 2). The amount

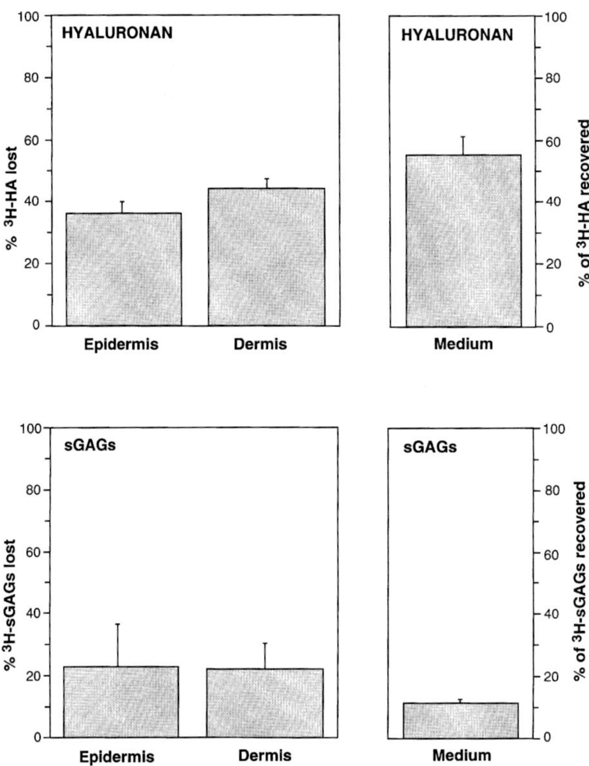

FIGURE 1 Disappearance rate of tissue GAGs in organ culture. The figures show the percentage of activity that disappeared from the tissues during the chase (means and SEM from five independent experiments). The amount of labeled GAGs recovered in the culture medium during the chase period is expressed per dermal tissue weight.

TABLE 2
HA Content in Skin after 5 Day Organ Culture[a]

	Hyaluronan (ng/mg tissue dry weight) 5-day culture
Epidermis	90 ± 10
Dermis	350 ± 40
Medium	3470 ± 390

[a] Culture media were changed twice during the period. After culture, epidermis and dermis were separated, dried, and digested with papain. HA was measured using biotinylated HABC as described in Materials and Methods. The data represent means and SEM from seven independent experiments. The HA recovered in the media during the 5-day culture period is calculated per dermal dry weight.

of HA recovered in the culture medium during the same period roughly corresponded to the amount lost from the dermal tissue (Table 2). The amount of HA appearing in the medium was highest during the first two culture days, then decreasing to a plateau where the amount of HA liberated daily into the medium corresponded approximately to that found in the dermal tissue (data not shown).

The disappearance rates of sGAGs from the epidermis and dermis appeared to be slower than those of HA (Figure 1). The amounts of sGAGs recovered in the culture medium during the chase were lower than that lost from dermal tissue, indicating that some of the PGs were taken up by the cells and degraded.

C. ION-EXCHANGE CHROMATOGRAPHY OF [³H]GLUCOSAMINE-LABELED MACROMOLECULES IN SKIN

[³H]glucosamine-labeled tissues were extracted with 4 M guanidinium chloride, desalted, and chromatographed on an ion-exchange column (Figure 2). Epidermal ³H-labeled material resolved into three distinct peaks during the NaCl gradient. Peak 1, eluting with low salt concentration, was not sensitive to *Streptomyces* hyaluronidase (Table 3) and had relatively low molecular weight (data not shown), suggesting that it contained glycoproteins. In dermal and medium samples, this peak was negligible (Figure 3). Peak II, eluting with a slightly higher salt concentration, was the major [³H]glucosamine-labeled peak in epidermal, dermal, and medium samples. Its elution position corresponded to that of standard HA, and it was susceptible to *Streptomyces* hyaluronidase identifying it as HA (Table 3). The elution position of peak III corresponded to that of standard HS and CS. In epidermal samples, 60% of the activity in peak III was susceptible to nitrous acid treatment, indicating that HS was the major sGAG synthesized in epidermis (Table 3). The rest of the activity was degraded with chondroitinase ABC. In dermal samples, 50 and 25% of peak III activity was degraded with chondroitinase ABC and nitrous acid, respectively, indicating that DS/CS and HS were the major dermal sGAGs synthesized in organ culture (Table 3).

D. MOLECULAR MASS DISTRIBUTION OF HA SYNTHESIZED IN ORGAN CULTURE

Peak II from MAP7+ chromatography was pooled for each sample and chromatographed on Sephacryl S-1000 (Figure 3). The ³H-labeled HA from both epidermis and dermis eluted in a large range from the site of high-molecular-mass standard HA (molecular mass of several million daltons) down to the elution position of standard CS (*K*av 0.8, molecular mass of 3×10^4). The mean molecular mass of epidermal HA was higher than that of dermal HA (Figure 3). The mean molecular mass of HA liberated into the culture medium during the labeling period was clearly smaller than that of dermis or epidermis (Figure 3).

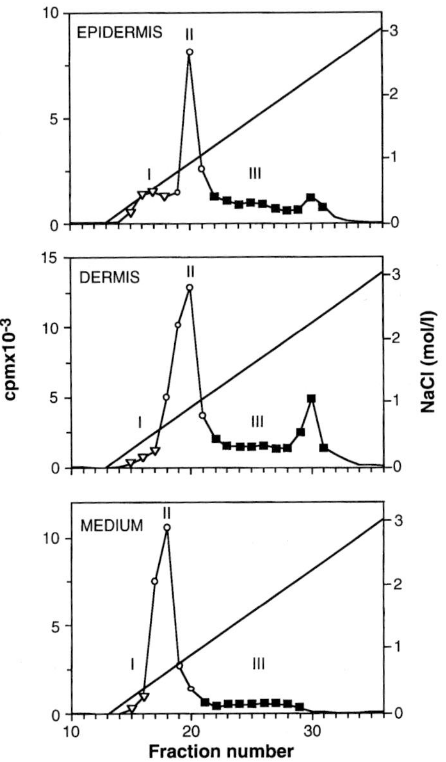

FIGURE 2 Anion-exchange chromatography of macromolecules labeled with [³H]glucosamine in whole skin organ culture. Fractions indicated by I (open triangles), II (open circles), and III (closed squares) were pooled for further analysis.

TABLE 3
Susceptibility of the Anion-Exchange Chromatography Peaks I, II, and III (Figure 2) to Degradation with Specific Enzymes and Nitrous Acid

	% of ³H activity degraded		
	Streptomyces hyaluronidase	Chondroitinase ABC	Nitrous acid
Epidermis			
Peak I	0		
Peak II	100		
Peak III		40	60
Dermis			
Peak I	13		
Peak II	100		
Peak III		50	25

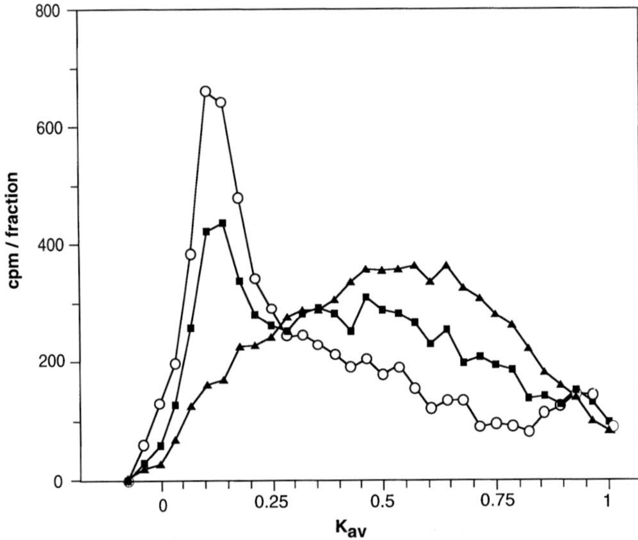

FIGURE 3   Molecular mass distribution of ³H-labeled HA in skin organ culture. Tissues were labeled with [³H]glucosamine for 24 h and HA extracted with 4 M guanidinium chloride from epidermis and dermis, purified with anion-exchange chromatography (Figure 2), and chromatographed on a Sephacryl S-1000 column. Open circles, epidermis; closed squares, dermis; closed triangles, medium.

E.   MOLECULAR MASS DISTRIBUTION OF PGs SYNTHESIZED IN ORGAN CULTURE

Skin explants were cultured in the medium containing ³⁵SO₄, and the PGs were extracted from the tissues using 4 M guanidinium chloride and 0.5% CHAPS. They were concentrated and partially purified with ethanol precipitation. Thereafter, samples were subjected to degradation with specific enzymes and analyzed by agarose gel electrophoresis.

Epidermal PGs migrated in agarose gel as a diffuse band between the 69-kDa protein standard and A1-PG (MW about 600 kDa) (Figure 4). The major band in epidermis had an apparent MW >200 kDa. Most of the activity was susceptible to heparitinase digestion, while chondroitinase ABC treatment caused disappearance of a small amount of activity in the fast moving front of the smear (Figure 4). While chondroitinase ABC digestion usually produced fragments so small that they were not seen in agarose gel electrophoresis, the fragments produced by heparitinase digestion were larger and comigrated with standard CS and the 69-kDa protein standard in agarose gel electrophoresis (Figure 4). Keratanase treatment showed no influence on the intensity or mobility of epidermal ³⁵S-labeled PGs.

³⁵SO₄-labeled material from dermis resolved in 0.7% agarose into three bands (Figure 4). The two smaller bands comigrated with CS standard and, the

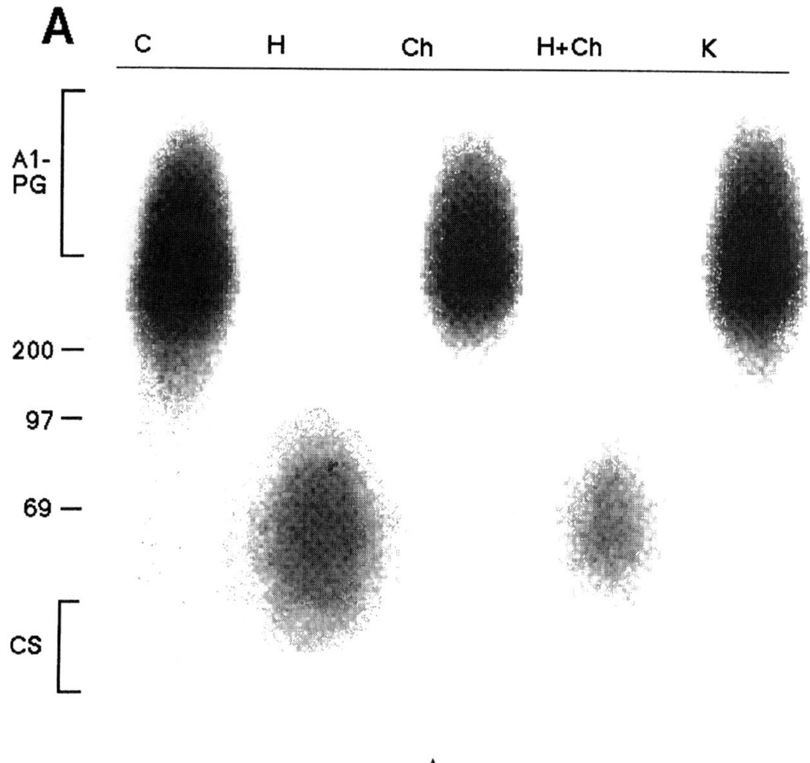

A

FIGURE 4 Horizontal electrophoresis, in 0.7% agarose, of $^{35}SO_4$-labeled epidermal (A) and dermal (B) PGs. Samples were electrophoresed without treatment (C), following digestion with heparitinase (H), chondroitinase ABC (Ch), keratanase (K), or with both heparitinase and chondroitinase ABC (H+Ch) prior to the electrophoresis. The positions of the ^{14}C-labeled protein standards (in kDa), A1-PG, and CS C are indicated.

69-kDa protein standard while the third was a smear with mobility ranging between the 97-kDa protein standard and A1-PG. The activity in this slowly moving smear could be resolved in three parts. Most of the activity comigrated with the 200-kDa protein standard and was susceptible to degradation with heparitinase (Figure 4). Activity comigrating with A1-PG was degraded with chondroitinase ABC treatment. The same was true for the activity in the fast moving front of the smear (corresponding to the mobility of the 97-kDa protein standard). When both heparitinase and chondroitinase ABC were used, a complete removal of all SO_4-labeled material with apparent MW >69 kDa was observed.

Chondroitinase ABC treatment caused a marked reduction in the amount of activity comigrating with the 69-kDa protein standard, while heparitinase caused a smaller reduction. The same was true for the activity comigrating with standard CS (Figure 4). Keratanase treatment did not influence the mobility or intensity of the dermal $^{35}SO_4$-labeled PGs.

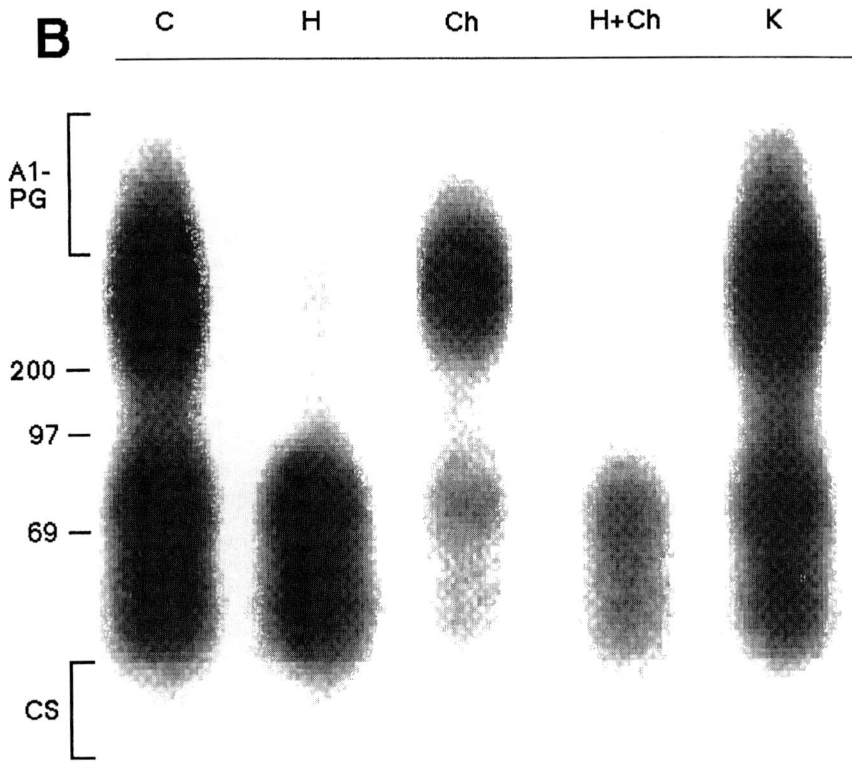

FIGURE 4B

IV. DISCUSSION

The present report demonstrates the use of organ culture of human skin to explore the metabolism of HA and PGs in epidermis and dermis under conditions that closely mimic *in vivo* conditions, but that are still chemically defined and controlled. A general tissue organization is maintained. Fibroblasts in their three-dimensional matrix with presumably normal matrix contacts show only limited cell proliferation.[62] In the epidermis, on the other hand, keratinocytes continuously proliferate forming new generations of cells committed to, and progressing through the stages of terminal differentiation.[50-52,62]

A. EXPRESSION OF HA AND PGS BY EPIDERMAL CELLS IN ORGAN CULTURE

In organ culture both epidermal and dermal cells actively synthesize HA and PGs (Table 1). The general spectrum of GAGs produced by human epidermis in organ culture was similar to that published for pig epidermis in organ culture,[53] and corresponds to their distribution in epidermis *in vivo*.[9]

Keratinocytes in cell culture also synthesize HA, HS, and CD/DS.[42,44-47] However, the relative proportions of different GAGs in keratinocyte cultures have given variable results in different reports.[42,44-47] The distribution is also different from that in organ culture models.[53,54] Some of the contradictory results may be attributed to differences in the keratinocyte proliferation rate and/or differentiation.[44,45,47] The fact that HS was the main sGAG synthesized by the epidermis in organ culture, while DS/CS formed a minor entity, is in accordance with the view that differentiating keratinocytes mainly express HS, whereas proliferating cells express more DS/CS.[47] In epidermis the differentiating cells form the majority of the cell population. The molecular heterogeneity of epidermal PGs revealed by electrophoretic analysis may reflect the whole range of maturational stages of keratinocytes in various layers of the epidermis. The identity of the various epidermal PGs synthesized in organ culture has not been shown. We have, however, preliminary data that a large, HS-substituted CD-44 is one of the major SO_4-labeled PGs in epidermis.[64]

B. METABOLISM OF EPIDERMAL HA AND DERMAL PGs IN ORGAN CULTURE

The epidermis *in vivo* is an isolated entity restricted basally by basal lamina and superficially by stratum corneum. These layers would most likely prevent the diffusion of HA and PGs. The same is true for epidermis cultured in organ culture. The HA pool in epidermis (mass HA/mass tissue) cultured for 5 days in organ culture closely corresponds to that *in vivo*.[63] In the present study we did not separate the proportions of extracellular and cellular HA/PGs in epidermis. King[53] reported that in pig epidermis cultured in organ culture, 80% of the total HA and about 50% of the sGAGS were extracellular and not bound to cells, indicating that under organ culture conditions epidermis contains an extracellular matrix composed of HA and PGs. In ordinary keratinocyte cell culture, extracellular HA is lost into the culture medium (about 50% of the total HA in culture).[44] The steady-state balance of synthesis and degradation of epidermal HA and PGs in organ culture makes this model ideal for metabolic studies of these compounds in epidermis. Furthermore, the maintenance of the normal extracellular matrix may be crucial when studying the influences of growth factors on keratinocyte HA and PG metabolism. PGs bind growth factors, thereby modulating their influences on cells.[3]

C. EXPRESSION OF HA AND PGs BY DERMAL CELLS IN ORGAN CULTURE

In vivo HA and DS comprise about 50 and 40% of dermal GAGs, respectively, while HS/heparin and CS form minor portions.[7,8] On the basis of the [³H]glucosamine incorporation rate, HA as the major GAG synthesized by dermal tissue in organ culture, while DS was the major sGAG (Tables 1 and 3). However, the rate of synthesis divided by amount is higher for HS than DS in dermis, suggesting faster metabolic turnover rate for HS-PGs compared to DS-PGs.

Most of the DS-PGs were relatively small, corresponding to the molecular mass of DS-PGII (decorin).[65] However, small amounts of large DS/CS-PGs (mobility in agarose gel electrophoresis comparable to aggrecan) were also synthesized. This corresponds to what has been published for cultured skin fibroblasts.[41,66] HS-PGs synthesized by the dermal tissue in organ culture were heterogeneous, probably representing several different species. Cultured skin fibroblasts have been reported to express seven different HS-PGs, some associated with the cells, some secreted to the medium.[39]

In addition to fibroblasts, dermal tissue contains various other cell types that express PGs, e.g., mast cells, endothelial cells, and the epithelial cells of skin appendages.[64,67] Furthermore, it is possible that even fibroblasts from papillary and reticular layers may differ in the spectrum of PGs they produce.[25] The diversity of the cell types thus obscures the origin of dermal PGs synthesized in organ culture. On the other hand, the organ culture model gives an overall view of the relative synthesis rates of various PGs in the whole tissue. This pattern can be compared with that obtained by chemical analyses performed on tissues. Cells in monolayer cultures are always more or less dedifferentiated and may thus synthesize PGs that are not expressed under normal *in vivo* conditions. The importance of cell–matrix contacts for PG gene expression has been shown in several cell culture systems where fibroblasts were grown on or inside artificial or endogenous collagen matrices.[38,66,68,69] However, these artificial models have given partly contradictory results. Fibroblasts cultured within an artificial collagen matrix show reduced expression of small DS-PG II (decorin)[38] compared to ordinary monolayer cultures, whereas fibroblasts embedded in their own endogenous matrix[66,69] or cultured on collagen matrix[68] expressed elevated quantities of decorin and versican-like molecules. The compositions of the artificial matrices are probably too simple and cannot provide the manyfold influences by which normal matrix regulates gene expression (tension on cells, changes in their shape, provision of cell matrix contacts, reservoirs of growth factors, proteolytic enzymes and the inhibitors, etc.). In organ culture the endogenous matrix with its different components presumably presents a full range of growth factors, and surrounds the cells, thereby exerting the wide range of extracellular signals regulating growth and differentiation.

D. METABOLISM OF DERMAL HA AND PGS IN ORGAN CULTURE

In fibroblast cultures either in or on collagen gels the amounts of PGs retained in the pericellular space are higher than in ordinary monolayer and correspond to the amounts found in the present organ culture model.[38,66,68,69]

The matrix binds DS-PGs (which are mainly secreted), thereby preventing their loss into the medium.[38] This binding may modulate the subsequent uptake and degradation of these PGs,[49] changing the proportion of PGs that are taken up and degraded by the cells. The overall disappearance rate of dermal PGs in organ culture corresponded to that reported for decorin in fibroblast cultures in collagen lattices.[38] The amounts of radioactive PGs recovered from the medium during a chase period were lower than the amounts disappearing from the tissue (Figure 1), suggesting local cellular uptake and degradation as a route of clearance for some of the dermal PGs.

The chase experiments and the analyses of total HA in dermal tissue and culture medium showed that most of the dermal HA diffuses into the culture medium (Figure 1). It has been suggested that also *in vivo* the vast majority of HA from dermis is drained into lymphatics to be further metabolized by lymph nodes and liver endothelial cells.[70] The rapid loss of HA from dermis in organ culture probably reflects the easier diffusion of HA into the medium. It also supports the idea that dermal HA for the most part is not tightly bound to cells or matrix molecules.[71] The reduction of the molecular mass of HA seems to precede and favor the diffusion out of the tissue as the average size of HA in the medium was smaller than that in the tissue (Figure 3). The factor(s) that causes the breakdown of dermal HA is unknown at present, but a similar size reduction in human skin fibroblast cultures was suggested to be caused by nonhyaluronidase activity present in the cultures.[72]

E.   REGULATION OF HA/PG METABOLISM IN SKIN ORGAN CULTURE

The organ culture model offers the possibility to modulate cell proliferation, differentiation, and migration by simple means thereby making correlations between these processes and HA/PG metabolism possible. For example, retinoic acid, a potent modulator of epidermal differentiation, enhances epidermal HA synthesis and leads to the accumulation of HA in the upper epidermal layers without significantly affecting dermal HA metabolism[54,63,73] or the metabolism of sGAGs.[54] Hydrocortisone opposes the actions of retinoic acid on keratinocyte differentiation and HA metabolism.[62] The data suggest that the regulation of HA expression in epidermis is tightly associated to keratinocyte differentiation. The experiments described in the present paper were done using chemically defined culture medium without any serum or serum-derived factors. Under these conditions, keratinocyte migration is minimal. By adding serum in culture medium it is possible to initiate keratinocyte migration onto the cut dermal surfaces thereby mimicking reepithelialization in wound healing.[74] By separating the migrating keratinocytes from the original epidermis, it is possible to follow the modulation of HA and PG metabolism during keratinocyte migration.

V. SUMMARY

We have studied the metabolism of HA and PGs in epidermal and dermal tissue compartments in whole skin organ culture model. The spectrum of GAGs produced by epidermis and dermis corresponds to their *in vivo* expression. In both tissues, HA is the major GAG synthesized, while HS is the main sGAG produced in epidermis and DS/CS in dermis. While the disappearance rate of HA is about 1 day in both epidermis and dermis, the route of clearance seems to differ. In epidermis a local degradation must occur, while most of dermal HA diffuses into the culture medium, mostly after depolymerization into medium-sized fragments. The disappearance rate of sulfated GAGs was slower than that of HA. Metabolic labeling and chase studies suggested the presence of several PG pools with slow and fast turnover rates in dermis. The present data give an overall view of various PGs expressed in skin organ culture. The expression and metabolism of individual PGs are possible using immunological purification techniques. Skin organ culture is technically easy to perform and maintains superior cell differentiation, facts that should make it an attractive alternative for ordinary cell culture methods in skin matrix research.

References

1. **Wright, T.N., Heinegård, D.K., and Hascall, V.C.,** Proteoglycans. Structure and function, in *Cell Biology of Extracellular Matrix*, 2nd ed., Hay, E.D., Ed., Plenum Press, New York, 1991, 45.
2. **Ruoslahti, E.,** Proteoglycans in cell regulation, *J. Biol. Chem.*, 264, 13369, 1989.
3. **Bernfield, M., Kokenyesi, R., Kato, M., Hinkes, M.T., et al.,** Biology of syndecans: a family of transmembrane heparan sulfate proteoglycans, *Annu. Rev. Cell Biol.*, 8, 365, 1992.
4. **Toole, B.P.,** Proteoglycans and hyaluronan in morphogenesis and differentiation, in *Cell Biology of Extracellular Matrix,* 2nd ed., Hay, E.D., Ed., Plenum Press, New York, 1991, 305.
5. **Laurent, T.C. and Fraser, J.R.E.,** Hyaluronan, *FASEB J.*, 6, 2397, 1992.
6. **Lis, J.M.J.v., Kruiswijk, T., Mager, W.H., and Kalsbeek, G.L.,** Glycosaminoglycans in human skin, *Br. J. Dermatol.*, 88, 355, 1973.
7. **Masuda, H., Shichijo, S., and Takeuchi, M.,** Comparative studies on the distribution of the glycopeptides and glycosaminoglycans in the human abdominal walls, *Int. J. Biochem.*, 8, 633, 1977.

8. **Poulsen, J.H. and Cramers, M.K.,** Determination of hyaluronic acid, dermatan sulphate, heparan sulphate and chondroitin 4/6 sulphate in human dermis, and a material of reference, *Scand. J. Clin. Lab. Invest.*, 42, 545, 1982.

9. **Mier, P.D. and Wood, M.,** Acid mucopolysaccharides of mammalian skin, *Br. J. Dermatol.*, 81, 528, 1969.

10. **Tammi, R.H., Hyyryläinen, A.M.H., Maibach, H.I., and Tammi, M.I.,** Ultrastructural localization of keratinocyte surface associated heparan sulphate proteoglycans in human epidermis, *Histochemistry*, 23, 1, 1987.

11. **Tammi, R., Ripellino, J.A., Margolis, R.U., and Tammi, M.,** Localization of epidermal hyaluronic acid using the hyaluronate binding region of cartilage proteoglycan as a specific probe, *J. Invest. Dermatol.*, 90, 412, 1988.

12. **Wang, C., Tammi, M., and Tammi, R.,** Distribution of hyaluronan and its CD44 receptor in the epithelia of human skin appendages, *Histochemistry*, 98, 105, 1992.

13. **Damle, S.P., Cöster, L., and Gregory, J.D.,** Proteodermatan sulfate isolated from pig skin, *J. Biol. Chem.*, 257, 5523, 1982.

14. **Fujii, N. and Nagai, Y.,** Isolation and characterization of proteodermatan sulfate from calf skin, *J. Biochem.*, 90, 1249, 1981.

15. **Matsunaga, E. and Shinkai, H.,** Two species of dermatan sulfate proteoglycans with different molecular sizes from newborn calf skin, *J. Invest. Dermatol.*, 87, 221, 1986.

16. **Miyamoto, I. and Nagase, S.,** Isolation and characterization of proteodermatan-sulfate from rat skin, *J. Biochem.*, 88, 1793, 1980.

17. **Nakamura, T., Matsunaga, E., and Shinkai, H.,** Isolation and some structural analyses of a proteodermatan sulphate from calf skin, *Biochem. J.*, 213, 289, 1983.

18. **Pearson, C.H., Winterbottom, N., Fackre, D.S., and Scott, P.G.,** The NH_2-terminal amino acid sequence of bovine skin proteodermatan sulfate, *J. Biol. Chem.*, 258, 15101, 1983.

19. **Choi, H.U., Johnson, T.L., Pal, S., Tang, L.-H., et al.,** Characterization of the dermatan sulfate proteoglycans, DS-PGI and DS-PGII, from bovine articular cartilage and skin isolated by octyl-Sepharose chromatography, *J. Biol. Chem.*, 264, 2876, 1989.

20. **Damle, S.P., Kieras, F.J., Tseng, W.-K., and Gregory, J.D.,** Isolation and characterization of proteochondroitin sulfate from pig skin, *J. Biol. Chem.*, 254, 1614, 1979.

21. **Habuchi, H., Kimata, K., and Suzuki, S.,** Changes in proteoglycan composition during development of rat skin. The occurrence in fetal skin of a chondroitin sulfate proteoglycan with high turnover rate, *J. Biol. Chem.*, 261, 1031, 1986.

22. **Poole, A.R., Webber, C., Pidoux, I., Choi, H., et al.,** Localization of a dermatan sulfate proteoglycan (DS-PG11) in cartilage and the presence of an immunologically related species in other tissues, *J. Histochem. Cytochem.*, 34, 619, 1986.

23. **Voss, B., Glössl, J., Cully, Z., and Kresse, H.,** Immunocytochemical investigation on the distribution of small chondroitin sulfate-dermatan sulfate proteoglycan in the human, *J. Histochem. Cytochem.*, 34, 1013, 1986.

24. **Yeo, T.-K., Brown, L., and Dvorak, H.F.,** Alterations in proteoglycan synthesis common to healing wounds and tumors, *Am J. Pathol.*, 138, 1437, 1991.

25. **Schönherr, E., Beavan, L.A., Hausser, H., Kresse, H., et al.,** Differences in decorin expression by papillary and reticular fibroblasts in vivo and in vitro, *Biochem. J.*, 290, 893, 1993.

26. **Bianco, P., Fisher, L.W., Young, M.F., Termine, J.D., et al.,** Expression and localization of two small proteoglycans biglycan and decorin in developing human skeletal and non-skeletal tissues, *J. Histochem. Cytochem.*, 38, 1549, 1990.

27. **Bosse, A., Schwarz, K., Vollmer, E., and Kresse, H.,** Divergent and co-localization of the two small proteoglycans decorin and proteoglycan-100 in human skeletal tissues and tumors, *J. Histochem. Cytochem.*, 41, 13, 1993.

28. **Zimmermann, D.R., Dours-Zimmermann, M.T., Schubert, M., and Bruckner-Tuderman, L.,** Versican is expressed in the proliferating zone in the epidermis and in association with the elastic network of the dermis, *J. Cell Biol.*, 124, 817, 1994.

29. **Yamagata, M., Shinomura, T., and Kimata, K.,** Tissue variation of two large chondroitin sulfate proteoglycans (PG-M/versican and PG-H/aggrecan) in chick embryos, *Anat. Embryol.* 187, 433, 1993.

30. **Hayashi, K., Hayashi, M., Jalkanen, M., Firestone, J.H., et al.,** Immunocytochemistry of cell surface heparan sulfate proteoglycan in mouse tissues. A light and electron microscopic study 1, *J. Histochem. Cytochem.*, 35, 1079, 1987.

31. **Inki, P., Larjava, H., Haapasalmi, K., Miettinen, H.M., et al.,** Expression of syndecan-1 is induced by differentiation and suppressed by malignant transformation of human keratinocytes, *Eur. J. Cell Biol.* 63, 43, 1994.

32. **David, G., Schueren, B.v.d., Marynen, P., Cassiman, J.-J., et al.,** Molecular cloning of amphiglycan, a novel integral membrane heparan sulfate proteoglycan expressed by epithelial and fibroblastic cells, *J. Cell Biol.*, 118, 961, 1992.

33. **Picker, L.J., Nakache, M., and Butcher, E.C.,** Monoclonal antibodies to human lymphocyte homing receptors define a novel class of adhesion molecules on diverse cell types, *J. Cell Biol.*, 109, 927, 1989.

34. **Murdoch, A.D., Liu, B., Schwarting, R., Tuan, R.S., et al.,** Widespread expression of perlecan protcoglycan in basement membranes and extracellular matrices of human tissues as detected by a novel monoclonal antibody against domain III and in situ hybridization, *J. Histochem. Cytochem.*, 42, 239, 1994.

35. **Carlstedt, I., Cöster, L., and Malmström, A.,** Isolation and characterization of dermatan sulphate and heparan sulphate proteoglycans from fibroblast culture, *Biochem. J.*, 197, 217, 1981.

36. **Johansson, S., Hedman, K., Kjellen, L., Christner, J., et al.,** Structure and interactions of proteoglycans in the extracellular matrix produced by cultured human fibroblasts, *Biochem. J.*, 232, 161, 1985.

37. **Heino, J., Kähäri, V.-M., Mauviel, A., and Krusius, T.,** Human recombinant intereleukin-1 regulates cellular mRNA levels of dermatan sulphate proteoglycan core protein, *Biochem. J.*, 252, 309, 1988.

38. **Greve, H., Blumberg, P., Schmidt, G., Schlumberger, W., et al.,** Influence of collagen lattice on the metabolism of small proteoglycan II by cultured fibroblasts, *Biochem. J.*, 269, 149, 1991.

39. **Schmidtchen, A., Carlstedt, I., Malmström, A., and Fransson, L.-Å.,** Inventory of human skin fibroblast proteoglycans, *Biochem. J.*, 265, 289, 1990.

40. **Westergren-Thorsson, G., Schmidtchen, A., Särnstrand, B., Fransson, L.-Å., et al.,** Transforming growth factor-β induces selective increase of proteoglycan production and changes in the copolymeric structure of dermatan sulphate in human skin fibroblasts, *Eur. J. Biochem.*, 205, 277, 1992.

41. **Breuer, B., Quentin, E., Cully, Z., Götte, M., et al.,** A novel large dermatan sulfate proteoglycan from human fibroblasts, *J. Biol. Chem.*, 266, 13224, 1991.

42. **Brown, K.W. and Parkinson, E.K.,** Glycoproteins and glycosaminoglycans of cultured normal human epidermal keratinocytes, *J. Cell. Sci.*, 61, 325, 1983.

43. **Konohana, A., Tajima, S., and Nishikawa, T.,** Glycosaminoglycan synthesis by cultured human epidermal cells, *Acta Derm. Venereol. (Stockholm)*, 66, 56, 1986.

44. **Lamberg, S.I., Yuspa, S.H., and Hascall, V.C.,** Synthesis of hyaluronic acid is decreased and synthesis of proteoglycans is increased when cultured mouse epidermal cells differentiate, *J. Invest. Dermatol.*, 86, 659, 1986.

45. **Piepkorn, M., Fleckman, P., Carney, H., and Linker, A.,** Glycosaminoglycan synthesis by proliferating and differentiated human keratinocytes in culture, *J. Invest. Dermatol.*, 88, 215, 1987.

46. **Roberts, G.P. and Jenner, L.,** Glycoproteins and glycosaminoglycans synthesized by human keratinocytes in culture, *Biochem. J.*, 212, 355, 1983.

47. **Rahemtulla, F., Moorer, C.M., and Wille, J.J., Jr.,** Biosynthesis of proteoglycans by proliferating and differentiating normal human keratinocytes cultured in serum-free medium, *J. Cell. Physiol.*, 140, 98, 1989.

48. **David, G. and Bernfield, M.R.,** Collagen reduces glycosaminoglycan degradation by cultured mammary epithelial cells: possible mechanism for basal lamina formation, *Proc. Natl. Acad. Sci. U.S.A.*, 76, 786, 1979.

49. **Schmidt, G., Hausser, H., and Kresse, H.,** Extracellular accumulation of small dermatan sulphate protoeglycan II by interference with a secretion-recapture pathway, *Biochem. J.*, 266, 591, 1990.

50. **Tammi, R. and Jansen, C.T.,** Effect of serum and oxygen tension on human skin organ culture: a histometric analysis, *Acta Dermatol. Venerol. (Stockholm)*, 60, 223, 1980.

51. **Tammi, R.,** A histometric and autoradiographic study of hydrocortisone action in cultured human epidermis, *Br. J. Dermatol.*, 105, 383, 1981.

52. **Tammi, R., and Santti, R.,** Ultrastructural study of hydrocortisone action in cultured human epidermis, *Br. J. Dermatol.*, 106, 65, 1982.

53. **King, I.A.,** Characterization of epidermal glycosaminoglycans synthesized in organ culture, *Biochim. Biophys. Acta*, 674, 87, 1981.

54. **Tammi, R. and Tammi, M.,** Influence of retinoic acid on the ultrastructure and hyaluronic acid synthesis of adult human epidermis in whole skin organ culture, *J. Cell. Physiol.*, 126, 389, 1986.

55. **Martikainen, A.-L., Tammi, M., and Tammi, R.,** Proteoglycans synthesized by adult human epidermis in whole skin organ culture, *J. Invest. Dermatol.*, 99, 623, 1992.

56. **Ågren, U., Tammi, R., and Tammi, M.,** A dot blot assay of metabolically radiolabeled hyaluronan, *Anal. Biochem.*, 217, 311, 1994.

57. **Saarni, H. and Tammi, M.,** A rapid method for separation and assay of radiolabeled mucopolysaccharides from cell culture medium, *Anal. Biochem.*, 81, 40, 1977.

58. **Tammi, R., Säämänen, A.-M., Maibach, H.I., and Tammi, M.,** Degradation of newly synthesized high molecular mass hyaluronan in the epidermal and dermal compartments of human skin in organ culture, *J. Invest. Dermatol.*, 97, 126, 1991.

59. **Lammi, M., and Tammi, M.,** Densitometric assay of nanogram quantities of proteoglycans precipitated on nitrocellulose membrane with safranin O, *Anal. Biochem.*, 168, 352, 1988.

60. **Lindahl, U., Bäckström, G., Jansson, L., and Hallén, A.,** Biosynthesis of heparin. II. Formation of sulfamino groups, *J. Biol. Chem.*, 248, 7234, 1973.

61. **Kongtawelert, P. and Ghosh, P.,** A method for the quantitation of hyaluronan (hyaluronic acid) in biological fluids using a labelled avidin-biotin technique, *Anal. Biochem.*, 185, 313, 1990.

62. **Ågren, U.M., Tammi, M., and Tammi, R.,** Hydrocortisone regulation of hyaluronan metabolism in human skin organ culture, *J. Cell. Physiol.*, 1995, in press.

63. **Tammi, R., Ripellino, J.A., Margolis, R.U., Maibach, H.I., et al.,** Hyaluronate accumulation in human epidermis treated with retinoic acid in skin organ culture, *J. Invest. Dermatol.*, 92, 326, 1989.

64. **Tuhkanen, A.-L., Tammi, M., Ågren, U., Törrönen, K., et al.,** unpublished.

65. **Björnsson, S.,** Size-dependent separation of proteoglycans by electrophoresis in gels of pure agarose, *Anal. Biochem.*, 210, 292, 1993.

66. **Qwarnström, E.E., Kinsella, M.G., MacFarlane, S.A., Page, R.C., et al.,** Modulation of proteoglycan metabolism by human fibroblasts maintained in an endogenous three dimensional matrix, *Eur. J. Cell Biol.*, 57, 101, 1992.

67. **Messenger, A.G., Elliott, K., Westgate, G.E., and Gibson, W.T.,** Distribution of extracellular matrix molecules in human hair follicles, *Ann. N.Y. Acad. Sci.*, 254, 1992.

68. **Gallagher, J.T., Gasiunas, N., and Schor, S.L.,** Synthesis of glycosaminoglycans by human skin fibroblasts cultured on collagen gels, *Biochem. J.*, 190, 243, 1980.

69. **Qwarnström, E.E., Järveläinen, H.T., Kinsella, M.G., Osteberg, C.O., et al.,** Interleukin-1β regulation of fibroblast proteoglycan synthesis involves a decrease in versican steady-state mRNA levels, *Biochem. J.*, 294, 613, 1993.

70. **Fraser, J.R.E. and Laurent, T.C.,** Turnover and metabolism of hyaluronan, in *Ciba Foundation Symposium*, 143, Whelan, J., Ed., Wiley, Chichester, 1989, 41.

71. **Reed, R.K., Laurent, T.C., and Taylor, A.E.,** Hyaluronan in prenodal lymph from skin: changes with lymph flow, *Am. J. Physiol.*, 259, H1097, 1990.

72. **Nakamura, T., Takagaki, K., Kubo, K., Morikawa, A., et al.,** Extracellular depolymerization of hyaluronic acid in cultured human skin fibroblasts, *Biochem. Biophys. Res. Commun.*, 172, 70, 1990.

73. **Tammi, R., Jansén, C.T., and Tammi, M.,** Effects of retinoic acid on adult human epidermis in whole skin organ culture, *Arch. Dermatol. Res.*, 277, 276, 1985.

74. **Tammi, R. and Jansén, C.T.,** Serum epidermal migration-stimulating factor: membrane ultrafiltration and physical stability characteristics, *Arch. Dermatol. Res.*, 267, 319, 1980.

8 *In Vivo/In Vitro* Assessments of Topical Hydrocortisone Availability: Correlation Between Blanching Assay and Laboratory Cell Experiments

Eric W. Smith and John M. Haigh

TABLE OF CONTENTS

I. INTRODUCTION

Topical corticosteroids are still the most widely used drugs in the treatment of dermatological conditions. Early corticosteroid dosage forms consisted of simple creams or ointments where more emphasis was placed on the potency of the drug molecule than on the intrinsic delivery potential of the vehicle.[1] More recently, the effect that the composition of the semisolid base has on the extent of drug delivery has been researched to a much greater extent.[2] These advances in the science of dosage form design have necessitated the refinement of precise and accurate methods for testing the drug delivery efficacies of the developed products. Obviously, the best method for the assessment of the effectiveness of corticosteroid formulations is in a therapeutic situation. Clinical trials, however, are fraught with methodological problems that make duplication of a trial impossible.[3] Alternatively, a number of pharmacological models[4] exist for this type of assessment, but it is often problematic to obtain correlation with the true dermatological conditions.

The human skin blanching assay is one of the most reliable and reproducible of the *in vivo* methods available for the assessment of topical corticosteroid formulations.[5] The skin whitening (blanching or vasoconstriction) side-effect that follows corticosteroid application was first utilized in 1962[6] as a measure of the percutaneous absorption of corticosteroids from topical formulations. Optimization of this initial procedure[7,8] has produced a reliable and precise bioassay methodology for the assessment of the efficacy of topical corticosteroid formulations.

One criticism of this assay has been the subjective nature of the observation procedure.[9,10] Although these points have repeatedly been addressed in the literature,[11,12] it has been suggested that it would be beneficial to have some *in vitro* penetration data to supplement *in vivo* observations, as this would strengthen the assessment of the topical equivalence of similar delivery formulations.[13,14]

With this objective in mind, a comparison of hydrocortisone release from two proprietary cream formulations was compared by *in vivo* and *in vitro* techniques to determine if any correlation could be established between the methodologies.

II. EXPERIMENTAL METHODOLOGY

The two commercial cream formulations used in this comparison were Cutaderm® cream (Scherag, South Africa) and an experimental cream formulation (Lennon, South Africa), each containing hydrocortisone at a concentration of 0.5%. Both formulations were assayed by a modified high-performance liquid chromatography technique[15] and were found to contain equivalent concentrations of hydrocortisone.

A. *IN VIVO* METHODOLOGY

The optimized methodology of the human skin blanching assay has been reported in detail previously.[5,7,8] Briefly, the assay used here employed the forearms of 12 subjects who had been screened for blanching response to topical corticosteroids. Four application patterns were used, each comprising 12 sites to which the formulations to be compared were randomly assigned so that each formulation was represented six times along the entire length of the forearm. The arm of each volunteer was randomly assigned one of the four patterns. Self-adhesive labels that had two 7 × 7-mm squares punched from their centers were used to demarcate the 12 forearm sites. Approximately 3 mg of each cream to be compared was extruded onto each site, according to the application patterns, from a small syringe, which had its needle cut to approximately 5 mm, and the formulation was uniformly spread using a glass rod. Once all the sites had been filled, the arm of each volunteer was occluded by covering each pair of sites with water-impervious tape. Only the occluded application mode was used here as this relatively weak corticosteroid, in low concentration, induces only slight blanching when tested without occlusion, which makes observation and interpretation of the results more difficult. The formulations, adhesive tapes, and dressings remained on the skin for 6 h, after which they were gently removed and any residual formulation washed from the skin. The first observation of skin blanching was made 1 h after product removal (7 h after application).

Three trained observers were used in this double-blind assessment to counteract the subjectivity of the methodology. The blanching response was independently assessed by the observers on nine occasions between 7 and 32 h after application. The intensity of blanching was graded on a 0 to 2 scale, 0 representing no blanching, 2 representing the most intense blanching expected, and 1 representing the intermediate grade. After the final observation the scores were decoded and the data from the three observers were summated for generation of a response profile (sum of the product of the frequency and observed score):

$$Summated\ score = \sum_{x=0}^{x=2} f(x)$$

The blanching response profiles generated in this way allow visual ranking of the formulations for comparative purposes. The qualitative data generated in this assay are statistically compared, for topical equivalence purposes, by nonparametric (χ^2) techniques.

B.   IN VITRO METHODOLOGY

The *in vitro* permeation cell methodology used in this study has been described previously.[16] Custom-made, glass diffusion cells of 10-ml volume were utilized in the vertical, "Franz" mode to measure the rate of hydrocortisone penetration from each cream formulation (packed into the donor chamber), through 1.65-cm^2 area full thickness hairless mouse skin, to a receptor phase of 10 ml purified isopropyl myristate maintained at 35°C and agitated by Teflon-coated bar stirrers at 600 rpm. The full techniques for membrane harvesting and preparation[17] and for high-performance liquid chromatographic analysis of the permeant in the receptor phase[18] have been reported in detail. In addition, the corticosteroid penetration performance of this cell design has been fully evaluated.[16]

III. RESULTS

A. *In Vivo* Assessment

The results of the human skin blanching assay are presented in Table 1 and are plotted in Figure 1 as the blanching response (summated score) versus the time after initial application of the formulations to the skin. These response profiles indicate that the blanching elicited by the experimental hydrocortisone formulation is greater in rank order than that of the Cutaderm® formulation. However, the profiles are similarly shaped and both peak at approximately 12 h. At face value, these drug delivery profiles do not appear to be significantly different, except possibly toward the end of the observation time where there is some divergence of the plots observed.

TABLE 1

In Vivo **Blanching Response Data for Commercial and Experimental Hydrocortisone Cream Formulations Applied in the Occluded Mode**

	Time after application (h)								
Score	**7**	**8**	**9**	**10**	**12**	**14**	**16**	**18**	**28**
				Cutaderm®					
0	144	86	80	66	59	71	113	143	181
1	70	128	129	137	123	124	96	73	35
2	2	2	7	13	34	21	7	0	0
Summated score	74	132	143	163	191	166	110	73	35
				Hydrocortisone					
0	129	97	81	63	60	68	105	130	174
1	84	113	124	127	113	108	96	76	40
2	3	6	11	26	43	40	15	10	2
Summated score	90	125	146	179	199	188	126	96	44

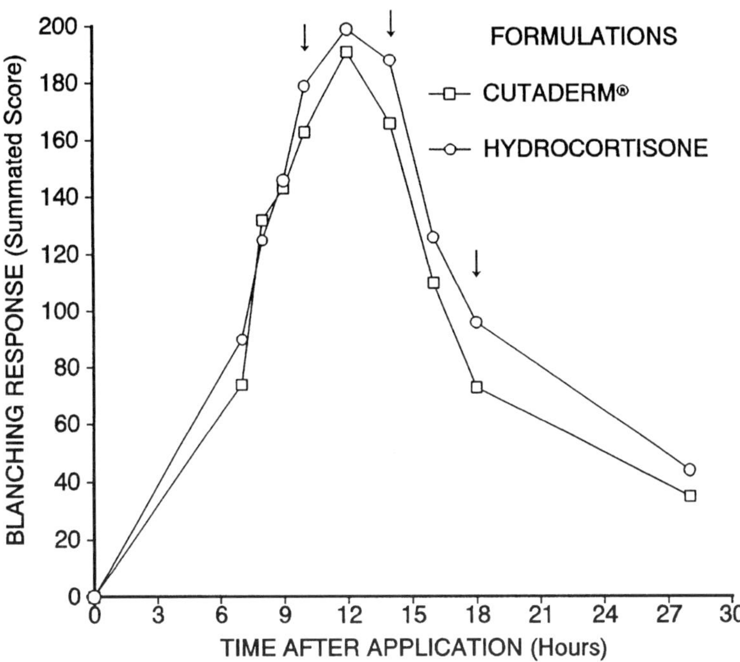

FIGURE 1   Human skin blanching response versus time after application profiles for commercial and experimental hydrocortisone cream formulations. Significant differences at the 95% level denoted by ↓.

B.   *In Vitro* Assessment

The rate of hydrocortisone permeation through full thickness hairless mouse skin under these experimental conditions is, essentially, equivalent from both cream formulations (Table 2). As can be seen from Figure 2, the two profiles intersect at various times and should show no statistically significant differences at any sampling time. The flux profiles for both creams appear approximately linear, especially during the first 24 h, with a slight increase in the hydrocortisone permeation rate observed at later sampling times. From these data it may be concluded that there is no difference in the rate of drug release or diffusion through hairless mouse skin from Cutaderm® cream and the experimental 0.5% hydrocortisone formulation.

The sensitivity of the high-performance liquid chromatographic technique employed in this assay allows receptor chamber hydrocortisone permeation flux as low as 0.5 μg cm^{-2} to be quantified. It is therefore possible to adequately monitor drug permeation with this cell system during the initial period of experimentation, the first quantifiable sample being withdrawn at 6 h. The lack

IN VITRO PERMEATION RESULTS

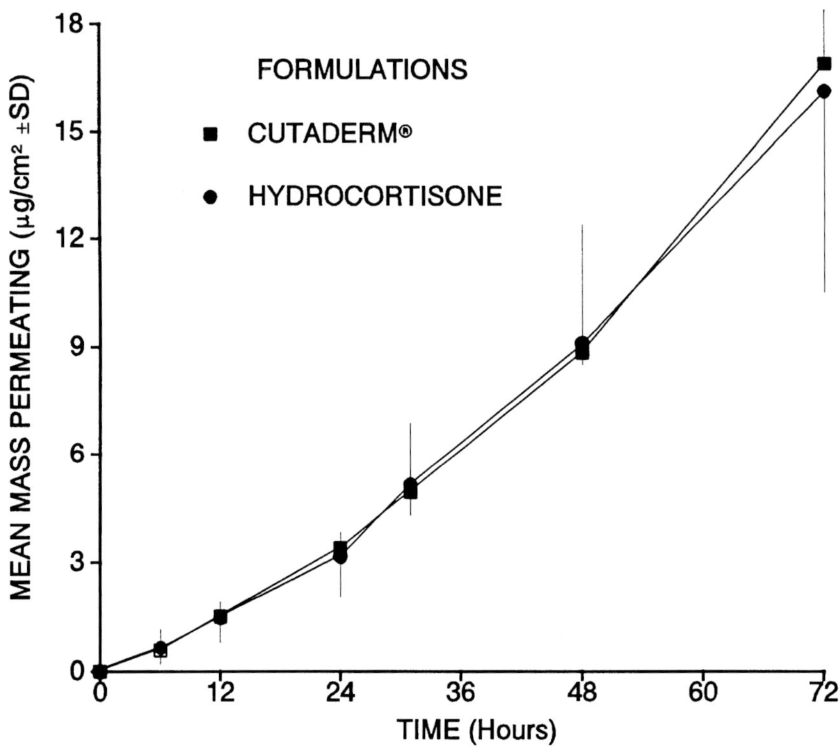

FIGURE 2   *In vitro* penetration of hydrocortisone through hairless mouse skin at 35°C from a commercial and an experimental cream formulation.

TABLE 2

In Vitro Drug Permeation Data for Commercial and Experimental Hydrocortisone Cream Formulations[a]

Time (h)	Mean drug mass permeating *in vitro*			
	Cutaderm®		Hydrocortisone	
6	0.59	(0.39)[b]	0.65	(0.49)
12	1.52	(0.39)	1.47	(0.67)
24	3.40	(0.44)	3.15	(1.10)
31	4.94	(0.66)	5.17	(1.70)
48	8.85	(0.35)	9.10	(3.29)
72	16.92	(1.48)	16.14	(5.61)

[a] Mean drug mass permeating hairless mouse skin at 35°C and 600 rpm agitation ($\mu g\ cm^{-2}$).

[b] Standard deviations of means in parentheses.

of an appreciable lag phase in the initial monitoring period would suggest that the permeant equilibrates rapidly with the barrier membrane and diffuses rapidly through it, into the receptor chamber. The laboratory data obtained here are also in congruity with results published for hydrocortisone permeation through synthetic membranes.[19] The latter study recorded approximately tenfold greater drug diffusion through the synthetic media than was observed for the hairless mouse skin in this study, a difference that would be expected, bearing in mind the general permeabilities of these two sets of membranes.[17]

C. STATISTICAL ANALYSIS

At face value, there appears to be no significant difference in the visually assessed blanching response profiles of the two formulations. The chi-squared (χ^2) distribution test was employed to determine the significance of the difference between the qualitative observations of the skin blanching intensity generated by the two formulations at each reading time. A two-by-two contingency table was used at each observation time that included the frequencies of the two highest grades of blanching observed (numbers of 1 and 2 grades). For the purposes of topical equivalence estimation, the period of maximal blanching (10 to 16 h) is considered most important. These data are depicted graphically in Figure 3 as the frequencies of each grade of blanching observed for each formulation over the peak response period. Significant differences in the

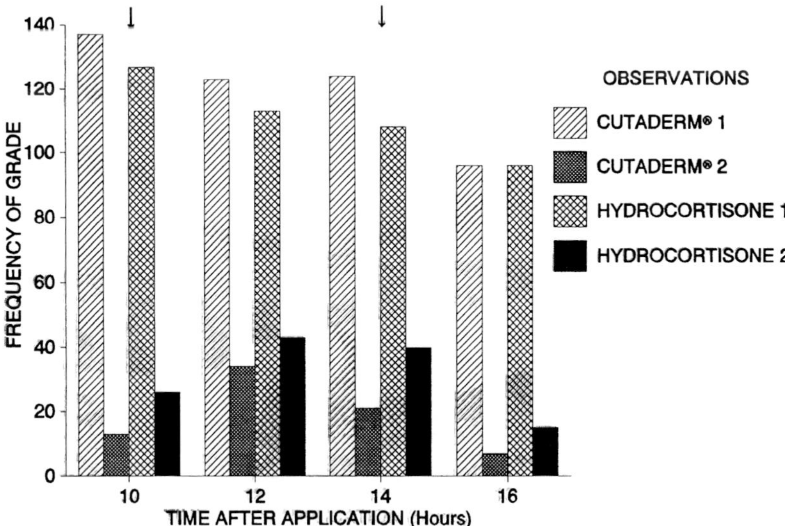

FIGURE 3 Frequencies of observed grades 1 and 2 for blanching generated by the commercial and experimental hydrocortisone cream formulations. Significant differences at the 95% level denoted by ↓.

data were found by this method at the 10- and 14-h observation times, with no significance in the difference being found at 12 and 16 h. Statistically, therefore, one cannot conclude with certainty that the degrees of blanching induced by the two formulations are sufficiently different over the peak observation times to ascribe inequivalence to their drug delivery patterns.

The quantitative data of the *in vitro* assessment are amenable to normal statistical treatment and may be compared by simple Student's *t*-distribution test methodology. As expected, there were no significant differences found, at the 95% level of significance, in the drug permeation from the two formulations at any sampling time. Statistically, therefore, the drug delivery from the two products is equivalent when using this *in vitro* methodology.

IV. CONCLUSIONS

Good correlation of the *in vitro* permeation data with the *in vivo* blanching results is, thus, demonstrated in this investigation. The diffusion cell permeation data monitored over 72 h indicates no significant differences in the drug release rates of the two hydrocortisone formulations, and, in congruity, the *in vivo* blanching data monitored over 28 h indicate that equivalent degrees of blanching are elicited by the two products. These methodologies, therefore, are complimentary to one another and may be used in combination for topical equivalence testing or quality control procedures. The availability of correlating systems such as this is becoming more important in topical drug absorption research, especially in regulatory affairs, for the registration of new products. It must be stressed, however, that the correlation observed here has been validated only for the two products compared in this study; there is no implication that this correlation will extend to the entire family of hydrocortisone products. With the diversity of topical delivery vehicles possible with technology available today, each formulation to be tested in this manner would require a fully validated *in vitro* protocol.[16]

References

1. **Smith, E.W.,** Do we need new and different glucocorticoids? A re-appraisal of the various congeners and potential alternatives, in *Current Problems in Dermatology*, Vol. 21, Korting, H.C. and Maibach, H.I., Eds., Karger, Basel, 1993, 1.
2. **Smith, E.W., Meyer, E., and Haigh, J.M.,** Blanching activities of betamethasone formulations. The effect of dosage form on topical drug availability, *Drug Res.*, 40, 618, 1990.

3. **Fredriksson, T., Lassus, A., and Salde, L.,** Reproducibility of clinical trials of topical glucocorticosteroids, *Int. J. Dermatol.*, 22, 536, 1983.

4. **Wend, H. and Frosch, P.J.,** *Clinico-pharmacological Models for the Assay of Topical Corticosteroids*, Karger, Basel, 1982.

5. **Haigh, J.M. and Kanfer, I.,** Assessment of topical corticosteroid preparations: the human skin blanching assay, *Int. J. Pharm.*, 19, 254, 1984.

6. **McKenzie, A.W. and Stoughton, R.B.,** Method for comparing percutaneous absorption of steroids, *Arch. Dermatol.*, 86, 608, 1962.

7. **Smith, E.W., Meyer, E., Haigh, J.M., and Maibach, H.I.,** The human skin blanching assay as an indicator of topical corticosteroid bioavailability and potency: an update, in *Percutaneous Absorption. Mechanisms — Methodology — Drug Delivery*, Bronaugh, R.L. and Maibach, H.I., Eds., Marcel Dekker, New York, 1989, 443.

8. **Smith, E.W., Meyer, E., and Haigh, J.M.,** The human skin blanching assay for topical corticosteroid bioavailability assessment, in *Topical Drug Bioavailability, Bioequivalence, and Penetration*, Shah, V. and Maibach, H.I., Eds., Plenum, New York, 1993, 155.

9. **Shah, V.P., Peck, C.C., and Skelly, J.P.,** Vasoconstriction-skin blanching assay for glucocorticoids: a critique, *Arch. Dermatol.*, 125, 1558, 1989.

10. **Conner, D.P., Zamani, K., Almirez, R.G., Millora, E., Nix, D., and Shah, V.P.,** Use of reflectance spectrophotometry in the human corticosteroid skin blanching assay, *J. Clin. Pharmacol.*, 33, 707, 1993.

11. **Smith, E.W., Meyer, E., and Haigh, J.M.,** Accuracy and reproducibility of the multiple-reading skin blanching assay, in *Topical Corticosteroids*, Maibach, H.I. and Surber, C., Eds., Karger, Basel, 1992, 65.

12. **Haigh, J.M. and Smith, E.W.,** Topical corticosteroid-induced skin blanching measurement: eye or instrument? *Arch. Dermatol.*, 127, 1065, 1991.

13. **Shah, V.P., Elkins, J., and Skelly, J.P.,** Relationship between in vivo skin blanching and in vitro release rate for betamethasone valerate creams, *J. Pharm. Sci.*, 81, 104, 1992.

14. **Shah, V.P., Behl, C.R., Flynn, G.L., Higuchi, W.I., and Schaefer, H.,** Principles and criteria in the development and optimization of topical therapeutic products, *Skin Pharmacol.*, 6, 72, 1993.

15. **Smith, E.W., Haigh, J.M., and Kanfer, I.,** A stability-indicating HPLC assay with on-line clean-up for betamethasone 17-valerate in topical dosage forms, *Int. J. Pharm.*, 27, 185, 1985.

16. **Smith, E.W. and Haigh, J.M.,** In vitro diffusion cell design and validation. II. Temperature, agitation and membrane effects on betamethasone 17-valerate permeation, *Acta Pharm. Nord.*, 4, 171, 1992.

17. **Haigh, J.M. and Smith, E.W.,** The selection and use of natural and synthetic membranes for in vitro diffusion experiments, *Eur. J. Pharm. Sci.*, 2, 331, 1994.

18. **Smith, E.W. and Haigh, J.M.,** In vitro diffusion cell design and validation. I. A stability-indicating high-performance liquid chromatographic assay for betamethasone 17-valerate in purified isopropyl myristate receptor phase, *Pharm. Res.*, 6, 431, 1989.

19. **Shah, V.P., Elkins, J., Lam, S.-Y., and Skelly, J.P.,** Determination of in vitro drug release from hydrocortisone creams, *Int. J. Pharm.*, 53, 53, 1989.

9 A New Model for Testing Phototoxicity of Skin Microorganisms

Jan Faergemann and Olle Larkö

TABLE OF CONTENTS

I. INTRODUCTION

Ultraviolet light has an inhibitory effect on many microorganisms.[1,2] However, microorganisms may also inhibit the growth of other microorganisms.[3,5] UVA may increase this effect for some microorganisms, but the chemicals responsible for this mechanism are, in most cases, still unknown. It is well known that many chemicals may have a lethal effect on microorganisms after UV irradiation *in vitro*. Certainly an exposure to longwave ultraviolet irradiation (PUVA) has a lethal effect on many microorganisms *in vitro*. However, in a study by Weissman and Nobel, no effect was seen on the cutaneous aerobic bacteria flora after PUVA treatment in patients with psoriasis.[6] Also, frequent sun bathing does not greatly affect the total number of microorganisms on the skin.[7]

The *Candida albicans* growth inhibition test, first described by Daniels,[8] for testing phototoxicity of various chemicals is still widely used. A modification of this method using shorter incubation time has been described by Wiskemann.[9] Later modifications on attempts to standardize the test procedure have also been described.[10-11] Mitchell, using a modification of Daniels' method for testing phototoxicity, has reported that some microorganisms may be phototoxic to other microorganisms.[3]

In this paper we describe in a new model the activity of ultraviolet light on microorganisms, and the phototoxic effect of microorganisms against other microorganisms, and the phototoxic killing effect of various microorganisms.

II. MATERIALS AND METHODS

A. THE EFFECT OF UV LIGHT ON HUMAN SKIN MICROORGANISMS

1. Text Organisms

Pityrosporum ovale ATCC 42132 and 44341, *Candida albicans* H29, *Staphylococcus epidermidis* CK, and *Staphylococcus aureus* 176 (kindly submitted by professor Lars Edebo, Department of Clinical Bacteriology, University of Gothenburg) were used. *P. ovale* was grown on a glucose–neopeptone–yeast extract agar medium containing olive oil (20 ml/l), Tween 80 (2 ml/l), and glycerol monosterate (2.5 g/l) earlier described.[12] *C. albicans* was grown on Sabouraud's agar medium and *S. epidermidis* and *S. aureus* on blood agar. The microorganisms were suspended in phosphate-buffered saline (PBS), pH 7.4, to give solutions containing 10^3 cells/ml.

2. Irradiation Technique

The microorganisms were irradiated with both UVB and UVA. Philips TL12 was used as a UVB source and Mutzhas UVA-SUN 2000 for UVA irradiation. For wavelength studies a monochromator, clinical photoirradiator (Applied photophysics) was employed. UVB irradiation was measured with an

International light (IL 1350 SED 240) probe and UVA with an International light (IL 1350 SED 015) probe. UV intensities for the monochromator were measured with a thermopile.

When a Philips TL12 UVB lamp was used 0.3 ml of the test organisms was spread with a bent glass rod on the appropriate medium and irradiated with doses from 65 to 900 mJ/cm². Control plates were not irradiated. Plates were then incubated at 37°C. *C. albicans*, *S. epidermidis*, and *S. aureaus* were read after 24 h and *P. ovale* after 3 days.

With Mutzhas UVA-SUN 2000, the experiment was done as described for UVB but the test plates were irradiated with doses from 25 to 27 J/cm². With a monochromator only *P. ovale* ATCC 44341 was tested. The result of irradiation at wavelengths of 300, 330, and 360 nm was studied. Plates were incubated on 5 to 8 different spots with one drop (20 µl) of the cell suspension. The spots were irradiated with increasing doses of light. A 300 ± 5 nm the doses were 10 to 1280 mJ/cm², at 330 ± 20 nm the doses were 2.5 to 40 J/cm², and at 360 ± 30 nm the doses were 5 to 80 J/cm². Control plates were not irradiated. After the experiment plates were incubated at 37°C for 3 days.

3. Statistics

The Student's *t* test for unpaired samples was used to compare the number of colony-forming units before and after irradiation.

B. THE PHOTOTOXIC INHIBITORY EFFECT AND PHOTOTOXIC KILLING EFFECT OF MICROORGANISMS

1. Test Organisms

S. aureus 176, *S. epidermidis* CK, *Pseudomonas aeruginosa*, *Escherichia coli*, *Streptococcus pyogenus* gr. A, *St. viridans*, *Propionibacterium acnes* (kindly provided by professor Lars Edebo, Department of Clinical Microbiology, University of Gothenburg), *C. albicans* H29, and *P. ovale* ATCC 42132 were used. The bacteria were grown on blood agar, *C. albicans* was grown on Sabouraud's agar, and *P. ovale* was grown on a glucose–neopeptone–yeast extract agar medium containing olive oil, Tween 80, and glycerol monostearate, as earlier described.[12] All microorganisms were incubated at 37°C.

2. Test Procedure

Philips TL 12 was used as a UBV source and Mutzhas UVA-SUN 3000 as a UVA source. UVB irradiation was measured with an IL 1350 SED 240 probe and UVA with an IL 1350 SED 015 probe.

C.   PHOTOTOXIC INHIBITORY EFFECT OF MICROORGANISMS

The target microorganism was suspended in PBS (0.3 ml, 10^5 cells/ml) and spread with a bented glass rod over DST-agar (Oxoid, Basingstoke, U.K.). A suspension (10^8 cells/ml) of the other microorganisms was placed in a well (5 mm) cut in the agar. When *P. ovale* was the target organism Tween 80 (2/ml) and glycerol monosterate (2.5 g/l) were added to the agar. With *P. acnes* blood agar was used because *P. acnes* was unable to grow on DST-agar. For *P. ovale* and *St. viridans*, an extract made from a 10^8 cells/ml suspension was also placed in the wells. A homogenate was obtained by grounding the cell suspension in a mortar for 5 min. The extract was then made by filtering the homogenate through a sterile Zeiss filter with a pore size of 0.45 μm. The extract contained mainly intracellular and soluble material. Plates were irradiated after 5 min prediffusion at room temperature, with a dose of 25 J/cm² UVA. *St. pyogenes* gr. A and *St. viridans* showed almost complete growth inhibition after an irradiation dose of 25 J/cm² and were therefore irradiated with a dose of 10 J/cm². Control plates were not irradiated. *P. ovale* was read after 3 days and the other microorganisms after 1 day.

D.   PHOTOTOXIC KILLING EFFECT OF MICROORGANISMS

In this experiment only *P. ovale* and *St. viridans* extracts were used. Before testing the phototoxic effect of these microorganisms and other microorganisms the direct effect of UVB irradiation on suspensions of all of the test organisms were studied. The effect of 250 mJ/cm² UVB was tested on all microorganisms. One milliliter suspension of the microorganisms in PBS was transferred to a quartz glass cuvette. One cuvette was irradiated with 250 mJ/cm² UVB and the other was not irradiated. Samples (0.1 ml) from a tenfold dilution of the microorganisms were then transferred to the appropriate culture medium and spread with a bent glass rod. *P. ovale* was inoculated onto the glucose–neopeptone–yeast extract agar medium with lipid supplements, *C. albicans* was transferred to a Sabouraud's agar medium, and the bacteria were incubated onto blood agar. The results for bacteria and *C. albicans* were read after 2 days and for *P. ovale* after 6 days of incubation at 37°C.

To test the phototoxic killing effect a suspension of the various microorganisms was suspended in the extracts made from *P. ovale* and *St. viridans*. One milliliter was transferred to quartz glass cuvettes. *St. pyogenes* gr. A was irradiated with 10 J/cm² of UVA and the other microorganisms with 25 J/cm² of UVA. Control cuvettes were not irradiated. The total contact time for both irradiated and nonirradiated organisms with the extract was 1 h. Samples (0.1 ml) from tenfold dilutions of the suspensions were then transferred to the appropriate growth culture medium.

III. RESULTS

A. THE EFFECT OF UVB ON SKIN MICROORGANISMS

The effects of UVB irradiation on *P. ovale*, *C. albicans*, *S. epidermidis*, and *S. aureus* are seen in Table 1. *P. ovale* was the most sensitive organism but also for *C. albicans* no growth was present at a dosage of 250 mJ/cm. *S. epidermidis* and *S. aureus* were more sensitive. For UVA only *P. ovale* was significantly reduced. Table 2 shows the effect of irradiation with monochromatic light. Again the reduction in number of cells was most pronounced at the shortest wavelength.

TABLE 1
Results of UVB and UVA Irradiation on *Pityrosporum ovale*, *Candida albicans*, *Staphylococcus epidermidis*, and *S. aureus*

	Growth inhibition	
Target organism	UVB (mJ cm^{-2})	UVA (J cm^{-2})
P. ovale		
ATCC 42132	250	75 (99% reduced)
ATCC 44341	250	
C. albicans H29	250	75 (20% reduced)
S. epidermidis CK	900	75 (75% reduced)
S. aureus 176	900 (97% red.)	75 (75% reduced)

Data adapted from Faergemann, J. and Larkö, O., *Acta Dermatol. Venereol. (Stockholm)*, 67, 69, 1987.

TABLE 2
Results of Monochromatic Irradiation on *Pityrosporum ovale*

Irradiation at	Growth inhibitory dose
300 nm (mJ cm^{-2})	640
330 nm (mJ cm^{-2})	160
360 nm (mJ cm^{-2})	80 (55% reduced)

Data adapted from Faergemann, J. and Larkö, O., *Acta Dermatol. Venereol. (Stockholm)*, 67, 69, 1987.

TABLE 3

The Growth Inhibitory Effect of *P. ovale* and *St. viridans* Cells on Skin Microorganisms

| | Growth inhibition (mm) | | | |
| | *St. viridans* | | *P. ovale* | |
Target organism	Control	UVA	Control	UVA
S. epidermidis	36	41	36	41
S. aureus	28	37	28	37
Ps. aeruginosa	11	25	20	29
E. coli	26	25	15	15
St. pyogenes gr. A.	35	35	0	0
St. viridans	0	0	0	0
C. albicans	0	0	0	0
Pr. acnes	0	0	0	0
P. ovale	0	22	0	0

TABLE 4

The Growth Inhibitory Effect of *P. ovale* and *St. viridans* Cell Extract on Skin Microorganisms

| | Growth inhibition (mm) | | | |
| | *St. viridans* | | *P. ovale* | |
Target organism	Control	UVA	Control	UVA
S. epidermidis	0	0	17	20
S. aureus	0	0	17	17
Ps. aeruginosa	0	12	10	12
E. coli	0	0	0	0
St. pyogenes gr. A	0	0	0	0
St. viridans	0	0	0	0
C. albicans	0	0	0	0
P. ovale	12	12	0	0

B. PHOTOTOXIC INHIBITORY EFFECT OF MICROORGANISMS

Only *P. ovale* and *St. viridans* had any inhibitory effect on the microorganisms tested. The inhibitory effect of *St. viridans* and *P. ovale* cells on the other microorganisms tested both in the dark and after UVB irradiation is shown in Table 3. The effects of extract from the same microorganisms are seen in Table 4.

TABLE 5
The Phototoxic Killing Effect of UVB on Microorganisms

Target organism	250 mJ/cm²	Control[a]
S. epidermidis	3.7×10^5	6.7×10^5
S. aureus	2.2×10^4	1.3×10^5
Ps. aeruginosa	6.0×10^3	1.8×10^5
E. coli	5.0×10^2	4.0×10^5
St. pyogenes gr. A	0	3.5×10^6
St. viridans	0	2.0×10^6
C. albicans	5.2×10^5	6.5×10^5
P. ovale	2.0×10^3	8.6×10^5

[a] Number of colonies.

The table is adapted from Faergemann, J. and Larkö, O., *Photodermatol. Photoimmunol. Photomed.*, 7, 35, 1990.

The whole-cell suspension had a stronger growth inhibitory effect than the extract indicating that the growth inhibitory substance is partly or in some cases exclusively bound to the cell membrane and/or cell wall.

C. PHOTOTOXIC KILLING EFFECT OF MICROORGANISMS

The direct killing effects of UVB on skin microorganisms arc shown in Table 5. *St. pyogenes* gr. A and *St. viridans* were totally killed after an irradiation dose of 250 mJ/cm² UVB. *E. coli* and *P. ovale* were also sensitive to UVB irradiation.

St. viridans and *P. ovale* cell extracts were not very effective in killing various skin microorganisms. The results after 1 h of contact with or without a pretreatment with UVA are shown in Table 6. The cell extracts did not increase the phototoxic activity of *St. viridans* and *P. ovale*.

IV. DISCUSSION

The *C. albicans* inhibition test for studying phototoxicity is still widely used both as originally described by Daniels[8] and with some modification.[9-11] The long irradiation time, variations in temperature, and the risk for the microorganisms to dry make this model less suitable. We have described a model for testing both the phototoxic inhibitory and phototoxic killing effect of microorganisms.[5] This model is also suitable for testing the phototoxic activity of various chemicals. In our model we use a defined inoculum size of

TABLE 6
The Phototoxic Killing Effect of *St. viridans* and *P. ovale* Extracts on Skin Microorganisms

| | Number of colonies | | | | | |
| | *St. viridans* | | *P. ovale* | | Control | |
Target organism	Dark	UVA	Dark	UVA	Dark	UVA
S. epidermidis	1.9×10^5	1.2×10^5	4.0×10^3	1.1×10^2	8.0×10^4	1.0×10^3
S. aureus	9.8×10^5	1.2×10^5	2.1×10^5	1.0×10^3	1.1×10^5	1.6×10^3
Ps. aeruginosa	2.8×10^6	1.1×10^6	4.5×10^6	2.8×10^6	2.1×10^6	1.1×10^6
E. coli	1.2×10^6	6.0×10^5	6.0×10^4	2.0×10^3	2.9×10^6	1.5×10^6
St. pyogenes gr. A	1.4×10^5	1.3×10^5	2.4×10^5	1.0×10^4	1.7×10^5	1.0×10^4

Data adapted from Faergemann, J. and Larkö, O., *Photodermatol. Photoimmunol. Photomed.*, 7, 35, 1990.

the microorganisms, a defined medium, the same method for applying the solutions or test microorganisms, the same prediffusion time, the same irradiation time, and the same incubation conditions. For testing the phototoxic activity of various chemicals *C. albicans* may be chosen because it has been used before, it is easy to maintain in culture, it grows easily when tested, and its pathogenicity is low.

Our results confirm earlier finding that some microorganisms may inhibit other microorganisms and this effect may be increased after UVB irradiation.[5] The difference in results between the effect of whole cells compared with extracts indicates that inhibitory substances are bound to the cell membrane and/or cell wall. Further studies are needed to finally isolate these substances.

The only microorganisms found to have inhibitory effects on other microorganisms were *P. ovale* and *St. viridans*. Mitchell has earlier shown that *P. ovale* may be phototoxic against some microorganisms.[3] He used blood agar for all microorganisms tested and the irradiation dose was low. In our model the inhibitory effects of *P. ovale* and *St. viridans* were markedly enhanced after irradiation, indicating that these two organisms have phototoxic effects on other microorganisms. No inhibitory effect of *Pr. acnes* was found. However, coproporphyrin III, present in *Pr. acnes*, is phototoxic, but the amount of coproporphyrin III present in *Pr. acnes* may be too low for a phototoxic reaction to develop. *P. ovale* is a member of the normal human cutaneous flora and *St. viridans* is a member of the normal flora in the mouth. The inhibitory effect of these two organisms against other microorganisms may play an important role in the whole ecology of human skin and mucous membranes of the mouth. It may also be an important defense mechanism against infections, but this needs to be further evaluated.

References

1. **Change, J.C.H., Ossoff, S.F., Lobe, D.C., et al.,** UV inactivation of pathogenic and indicator microorganisms, *Appl. Environ. Microbiol.,* 49, 1361, 1985.
2. **Faergemann, J. and Larkö, O.,** The effect of UV-light on human skin microorganisms, *Acta Dermatol. Venereol. (Stockholm),* 67, 69, 1987.
3. **Mitchell, J.C.,** Cutaneous floral and microfloral phototoxicity, *Contact Dermat.,* 8, 75, 1982.
4. **Faergemann, J. and Larkö, O.,** Phototoxicity of skin microorganisms tested with a new model, *Arch. Dermatol. Res.,* 280, 168, 1988.
5. **Faergemann, J. and Larkö, O.,** The phototoxic inhibitory effect and phototoxic killing effect of micro-organisms, *Photodermatol. Photoimmunol. Photomed.,* 7, 35, 1990.
6. **Weissman, A. and Nobel, W.C.,** Photochemotherapy of psoriasis: effects on bacteria and surface lipids in uninvolved skin, *Br. J. Dermatol.,* 102, 185, 1980.
7. **Gerber, D., Mathews-Roth, M., Fahlund, C., Hummell, D., and Rosner, B.,** Effect of frequent sun exposure on bacterial colonization of skin, *Int. J. Dermatol.,* 18, 571, 1979.
8. **Daniels, F.,** A simple microbiological method for demonstrating phototoxic compounds, *J. Invest. Dermatol.,* 44, 259, 1965.
9. **Wiskemann, A.,** Phototoxizitäts-Bestimmungen an Paramecien und *Candida albicans,* in *Photochemotherapie-Grundlagen, Technik und Nebenwirkungen,* Jung, E.G., Ed., Schattauer, Stuttgart, 1976, 19.
10. **Kavli, G. and Volden, G.,** The *Candida* test for phototoxicity, *Photodermatology,* 1, 204, 1984.
11. **Knudsen, E.A.,** The *Candida* phototoxicity test. The sensitivity of different strains and aspects of *Candida,* standardization attempts and analysis of the dose-response curves for 5- and 8-methoxypsoralen, *Photodermatology,* 2, 80, 1985.
12. **Faergemann, J. and Fredriksson, T.,** Experimental infections in rabbits and humans with *Pityrosporum orbiculare* and *P. ovale, J. Invest. Dermatol.,* 77, 314, 1981.

10 Drug and Cosmetics Evaluations with Skin Strippings

Gérald E. Piérard and Claudine Piérard-Franchimont

TABLE OF CONTENTS

I.   INTRODUCTION

The stratum corneum is a thin continuous membrane formed by tightly stacked corneocytes separated by a multilamellar organization of lipids. This structure undergoes a maturation process during its upward move to the surface of the skin with progressive reduction in intercorneocyte cohesion, ending with physiologic desquamation. The deep cohesive part of the stratum corneum is called stratum compactum, and the loose superficial part is referred to as stratum disjunctum.

Considering the constant renewal of the epidermis and the horny layer as its end product, the structure of the stratum corneum may be viewed as a recollection of the past history of the life of the epidermis. Alterations in the organization of corneocytes, parakeratosis, and many other aspects are therefore clues for the intervention of previous physiopathological disturbances.

Drugs may affect the stratum corneum and the functions that it subserves by the indirect route through modulations in the physiology of the stratum Malpighi. Drugs and cosmetics may also act directly on the structure of the stratum corneum. No proper pharmacology in the stratum corneum can exist without adequate techniques for the objective and quantitative recording of the effects of the substances tested.[1] In this chapter, reference will be made to the use of some of the techniques developed in our group in the field of pharmacology using skin strippings.

II.   METHODS USED FOR STRATUM CORNEUM COLLECTION

Obtaining samples of stratum corneum by skin scraping is a time-honored procedure. The collected material is obviously torn out and fragmented. The surface topography and cellular details are not appreciated with such sampling. Stripping is the classical way to collect the superficial part of the stratum corneum. The use of casual sticky tapes is not reliable for quantitative assessments because the adhesion to the stratum corneum is not controlled. An improved method using a liquid, nondrying, pressure-sensitive adhesive on a glass slide allows a better appreciation of the structure of the stratum corneum.[2] At the present time, the standard operating procedures in our laboratory include two other methods, cyanoacrylate skin surface stripping (CSSS) and stripping with adhesive-coated discs (SACD).

The method of CSSS was launched 25 years ago[3] and other groups have in time expanded and refined its use.[4-8] A droplet of cyanoacrylate glue is deposited onto a glass slide or on a sheet of clear polyethylene (Melinex O, ICI plastic division). This material is pressed against the surface of the skin for at least 30 s. In the presence of moisture, the cyanoacrylate polymerizes and adheres to the stratum corneum. The material is gently lifted and peeled from the skin.

The sampling device for SACD is a crystal clear adhesive-coated disc (D-Squame[R], Cuderm Corporation, Dallas, TX or Corneodisc L'Oréal, Paris, France) that provides the required rigidity and adhesion to uniformly sample a fixed area of skin surface. After peeling off the protective seal, the disc is applied to the skin surface by using a gauge spring dynamometer to ensure a standardized pressure.[9] Both the pressure and the time of application of the disc influence the amount of stratum corneum removed. A short-time (5 s) application of the disc removes less stratum corneum than a long-time (1 h) application. This is due to the occlusion modifying the hydration and cohesion of corneocytes.

If greasy products have been applied to the skin before sampling, the adhesion of the discs to the stratum corneum is impaired yielding unreliable data. In such experimental conditions, sampling errors are reduced by delipidizing the skin with ether:acetone (1:1).

III. BIOMETROLOGIC AND ANALYTIC EVALUATIONS OF SKIN STRIPPINGS

Skin strippings, including CSSS and SACD, are used as noninvasive diagnostic methods useful in many skin conditions.[4,6-8,10] They are also tools in combination with analytic procedures for the evaluation of the pharmacokinetics and efficacy of drugs. The assessment of the interaction between skin and cosmetics and toiletries also benefits from such approaches. In these respects, skin strippings are used in four main ways:

- evaluation of the structure of the stratum corneum
- evaluation of the dynamics of the stratum corneum renewal
- *in vivo* evaluation of antimicrobial
- use of CSSS as a substrate for bioassay

The challenge of analytic evaluations of skin strippings is to quantify some aspects of the structure and functions of the stratum corneum. Analytic morphology includes techniques that provide quantitative and statistically evaluable data from image acquisition. Image analysis of the aspect of samples seen under the microscope and optical profilometry of the surface of the samples are two of the classical methods of analytic morphology that are conveniently applied to skin strippings. Other biometrologic approaches have been applied, including reflectance colorimetry and light transmission evaluations. The collected stratum corneum on SACD can also be analyzed for the presence of drugs or other constituents. The 6-mm D-Squame discs allow SACD to be easily placed in the well of a microanalysis plate. The other sizes (13 and 22 mm) are best suited for analytic morphology and reflectance colorimetry.

IV. STRUCTURE OF THE STRATUM CORNEUM MODIFIED *IN VIVO* BY DRUGS AND COSMETICS

A. XEROSIS

Some drugs and cosmetics alter the pathophysiology of desquamation and scaling. Normal desquamation takes place imperceptibly as single corneocytes are released at the skin surface. If the intracorneal cohesion fails to drop, the corneocytes stick together and separate as large clumps, giving rise to scaling. The feel by the sensing finger and the appearance by the reflection of the incident light lead to xerosis often being called rough dry skin. In fact, the concept of dry skin is probably not so accurate and is a matter of controversy.[1,11,12] Its definition and limits are idiosyncratic, varying according to different local traditions in the laboratories. Emollients and keratolytics (which are better named squamolytic agents) may be evaluated by skin strippings.

A semiquantitative grading of xerosis may be achieved by CSSS stained for 3 min with a solution of basic fuchsin and toluidine blue (Polychrome Multiple Stain, PMS, Delasco).[6,8,9] The specimens are graded according to the following scheme:

0: normal stratum corneum
1: hyperkeratosis of the first and second lines and/or the appendageal orifices
2: hyperkeratosis covering less than 30% of the plateaus
3: hyperkeratosis covering more than 30% of the plateaus
4: diffuse confluent scales
5: thick, uneven scales covering the entire surface, obliterating the skin surface furrows

It is also possible to measure the size of corneocytes and to detect the presence of parakeratosis and other cells on such specimens. Optical profilometry[13,14] may be applied to the CSSS. There is a positive linear correlation between the values of skin roughness (Ra and Rz) gained on CSSS and on skin replicas with Silflo[R]. The advantage of using CSSS over replicas is evident in scaly disorders when scales may adhere to the replica and obscure the data of optical profilometry.

Quantitative assessment of xerosis has recently been updated by the development of analytic evaluations of SACD. The attenuation of optical transmission of light has been used to measure the quantity of scales on the discs.[15] A variant of that method consists in measuring the color (Chroma Meter CR200 Minolta) of the unstained sample deposited onto a red-colored reference plate. Thick scales decrease the L^* and a^* values of reflectance colorimetry. Image analysis of SACD can also be used for the quantification of scaling disorders.[16,17] The parameters of importance are the area (A) occupied by scales and their thickness evaluated on a 5-level gray scale. The percentage of scales (Tn)

is calculated in relation to each thickness level (n). A desquamation index (DI) is yielded according to

$$DI = \frac{2A + \sum\left[Tn(n-1)\right]}{6}$$

Squamometry is still another objective evaluation of xerosis and some dermatitis.[9,18,19] SACD are stained for 1 min by dropping the staining solution PMS over the surface, followed by gentle rinsing with tap water. The stained SACD is placed over a hole cut out of a glass slide, which is then placed onto a Minolta reference plate for white color. The value of Chroma C^* measured by reflectance colorimetry correlates with the amount of removed scales and with the clinical rating of xerosis.[9] This technique has also been applied to evaluate inflammatory diseases such as atopic dermatitis[18] and to quantify the effect of antidandruff shampoos.[19]

B. FOLLICULAR CASTS AND MICROCOMEDONES

Follicular casts and microcomedones may be conveniently studied on CSSS.[5,8,20-23] The material sampled from the upper portion of the follicular ducts reflects the balance between comedo formation and lysis.

Analytic methods for evaluating the amount of follicular casts concern image analysis when illuminating the specimen with white light, polarized light, or fluorescent light.[22-24] The use of fluorescence for evaluating the presence of porphyrins produced by *Propionibacterium acnes* in follicles may prove to be difficult to interpret in relation to drugs and cosmetics. In fact, some products emit fluorescence by themselves and others display a quenching effect by absorption of porphyrin fluorescence. Moreover, fluorescence is not always limited to comedones and stratum corneum creases may fluoresce as well.

C. EXAMPLES OF THE USE OF SKIN STRIPPINGS FOR EVALUATING *IN VIVO* THE EFFECTS OF DRUGS AND COSMETICS ON THE STRUCTURE OF THE STRATUM CORNEUM AND ON THE AMOUNT OF FOLLICULAR CASTS

The efficacy of squamolytic agents and emollients is conveniently evaluated on skin strippings taken during a clinical trial using the regression method of Kligman.[25] In general, it is appropriate to take samples at entry in the study, as well as after 2 and 4 weeks of treatment, which is then stopped.

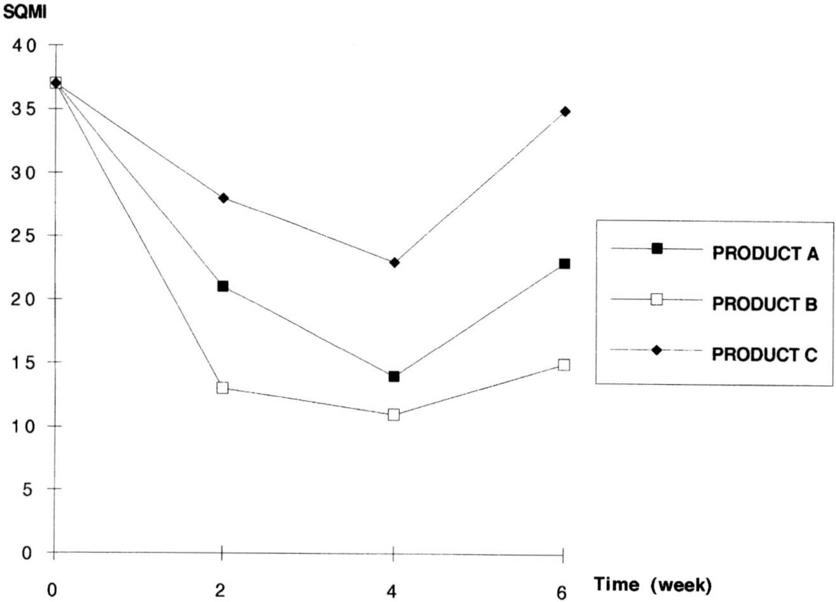

FIGURE 1 Evolution in the values of squamometry (Chroma C^* of SACD) showing different kinetics of improvement of xerosis during treatment (weeks 0 to 4) with three products and the following regression phase (weeks 4 to 6).

After a 2-week period without treatment corresponding to the regression phase, late samples are collected. Xerosis is rated on CSSS. Squamometry and desquamation index show the kinetics of improvement followed by the post-therapeutic regression (Figure 1).

The efficacy of corticosteroids in spongiotic dermatoses may also be evaluated by the same approaches. CSSS is used for rating the presence of serum, parakeratotic cells, and inflammatory cells in the stratum corneum. Squamometry values decrease when the inflammatory signs disappear in the stratum corneum.[18]

Xerosis of old age and wrinkling may be evaluated on CSSS.[6] In particular, the wrinkling process and its correction by some drugs and cosmetics is conveniently substantiated by image analysis and optical profilometry.

CSSS are used to predict the comedogenic risk linked to products, and to quantify the comedolytic activity of drugs and cosmetics.[21-23] Both the number and the size of follicular casts are influenced by some treatments. The concentration of the active compound governs the biologic effect on microcomedones.

V.   EVALUATION OF THE DYNAMICS OF THE STRATUM CORNEUM RENEWAL

The dansyl chloride test is generally accepted as a noninvasive attempt at evaluating stratum corneum turnover.[26] The time of fluorescence extinction depends on both the rate of transit of corneocytes through the horny layer and the thickness of that layer. This test proves to be difficult to interpret due to the uneven fade-out of fluorescence at the skin surface. An improvement in the method was introduced by replacing dansyl chloride by dihydroxyacetone.[27] Further refinements are gained by taking CSSS from the test sites at a predetermined time of experiment.[8,28] At day 10, the specimens are examined under a fluorescent light microscope because both dansyl chloride and DHA browning agents are fluorescent. Image analysis applied to such pictures allows the quantification of the nonfluorescent vs. fluorescent areas of the stratum corneum. This ratio is an indicator of the rate of stratum corneum turnover. However, both the dansyl chloride test and the dihydroxyacetone test are dramatically influenced by the use of cleansing agents and skin care products.[27-29] The extraction of the dyes from the stratum corneum by these products may be used to predict irritancy.[29]

An inverse relationship has been established between the size of corneocytes and the speed of epidermal turnover.[30] This aspect may be studied on skin strippings.

Another facet of the dynamics of the stratum corneum is gained by squamometry. For instance, the kinetics of squamometry after acute photodamage shows a complex response of the epidermis with early (±48 h) and late (±10 days) changes (Figure 2).

VI. *IN VIVO* EVALUATION OF ANTIMICROBIAL

CSSS are well suited to evaluate the efficacy of antimicrobial in superficial skin infections. Both the density and the aspect of fungi are altered by antifungals (Figure 3). By using vital staining with neutral red, it is possible to distinguish dead fungi from viable forms.[31]

The load of *Malassezia ovalis* in squames of dandruff and seborrheic dermatitis may be estimated on SACD (Figure 3). The efficacy of shampoos can be compared by combining squamometry and semiquantitative evaluations of the number of yeasts.[19,32]

VII. BIOASSAY USING CSSS

The stratum corneum collected by CSSS may be used as a substrate for *ex vivo* bioassay. We have applied it for four main predictive purposes aiming at evaluating

- the efficacy and lingering effect of antifungals
- the irritancy of surfactants and the protection by skin barrier creams
- the contribution of the stratum corneum in the sensitive skin problem
- the potency of squamolytic agents

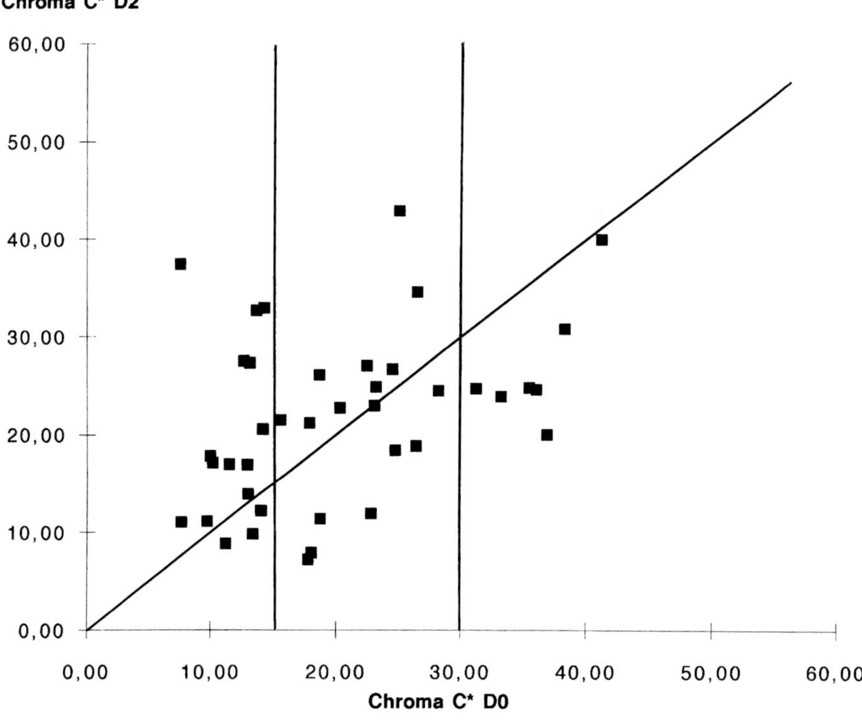

FIGURE 2   Squamometry (SQMI) before (D0) and 48 h (D2) after an acute photodamage.
Normal skin (Chroma C^* D0 <15) shows an increase in SQMI while xerotic skin (Chroma C^*
D0 >30) responds with a decrease in SQMI.

A.   EFFICACY AND LINGERING EFFECT OF ANTIFUNGALS

The evaluation of the antifungal activity in the stratum corneum is possible
when skin strippings are taken from the normal skin of volunteers under
therapy. The specimens are extracted for drug quantification or bioactivity
against specific fungal growth *in vitro*.[33-35] According to the experimental
protocol, the results are interpreted in terms of pharmacokinetics and
pharmacodynamics.

The bioactivity of topical and oral antifungals is also evaluated using
cultures of fungi on skin strippings. In fact, some fungi of medical importance
grow *in vitro* on skin strippings. This is possible on tape strippings,[36,37] but a
better method relies on the use of CSSS.[38,39] When samples are taken from
volunteers under antifungal therapy, the subsequent growth of fungi on CSSS
is inhibited. Morphometric evaluations of the fungal growth extent allow
quantification of the bioactivity and the kinetics of incorporation and elimination of the drug in the stratum corneum.[40-44]

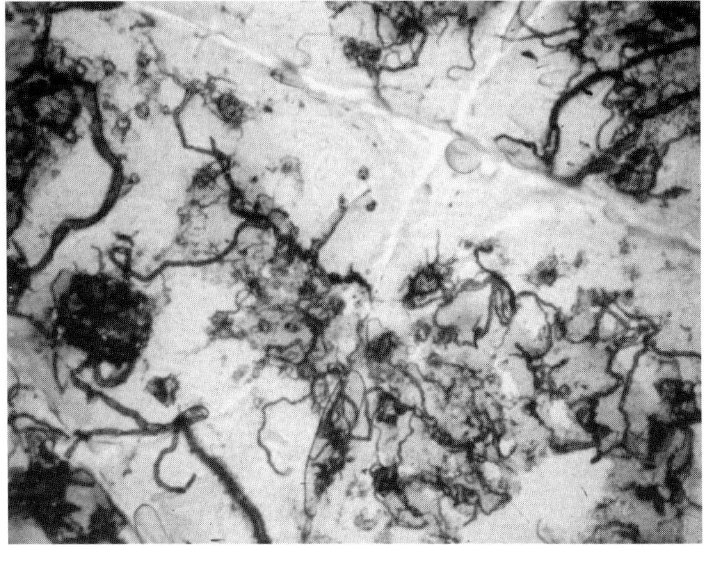

A

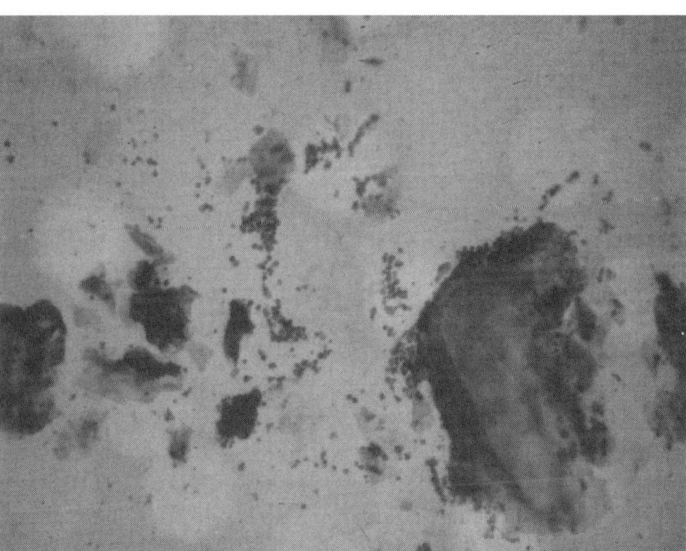

B

FIGURE 3   (A) Altered shape of fungi during treatment with oral triazole. (B) Dandruff with many *Malassezia ovalis.*

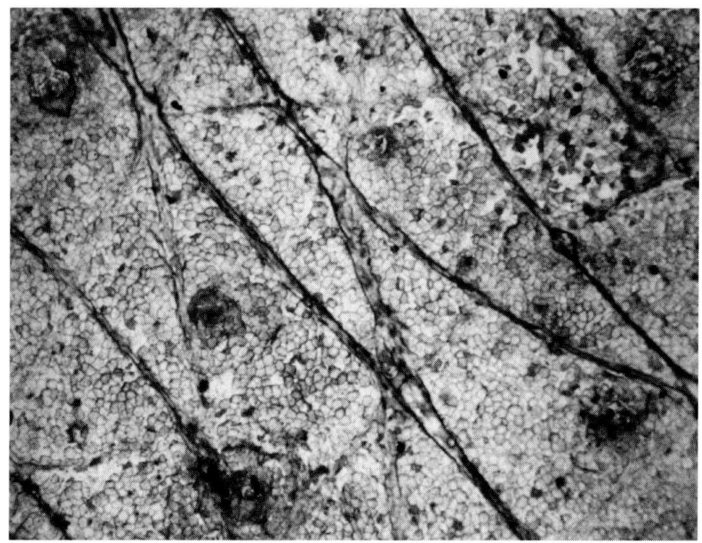

A

FIGURE 4 Microscopic aspect of a corneosurfametry test with mild (A) and harsh (B) surfactants. Intensity of staining with PMS is related to the irritant properties of the products. Limits of corneocytes and dotted structures resembling corneosomes are strongly stained (C).

B. CORNEOSURFAMETRY, A PREDICTIVE ASSAY FOR SURFACTANT IRRITANCY AND EFFICACY OF SKIN BARRIER CREAMS

Corneosurfametry is a noninvasive quantitative test rating the interaction between surfactants and human stratum corneum.[45] It may be used as a predictive irritancy test that compares favorably with classical *in vitro* tests.[46]

Corneosurfametry entails collection of CSSS and a predetermined contact time of the samples with diluted surfactants. A 2-h experiment is usually convenient to disclose differences among products. The usual range of test concentrations for proprietary cleansing skin care products is 5 to 10%, but higher dilutions may be needed for harsh products. Each surfactant solution is sprayed uniformly on 15 to 20 CSSS placed in a slide tray, which is then covered by lids. At completion of the predetermined evaluation time, CSSS are thoroughly rinsed with tap water, dried, and stained for 3 min with basic fuchsin–toluidine blue dye solution (PMS). After removing the excess of dyes and abundant rinsing with water, reflectance colorimetry is applied to the dry samples. Both L^* and Chroma C^* are recorded. To be valid for interpretations, the L^* value should be over 45 and superior to that of Chroma C^*.[45,47] Data out of range indicate that the concentration of surfactant is too high. For instance, the maximum concentration for testing sodium lauryl sulfate (SLS) is 2%.

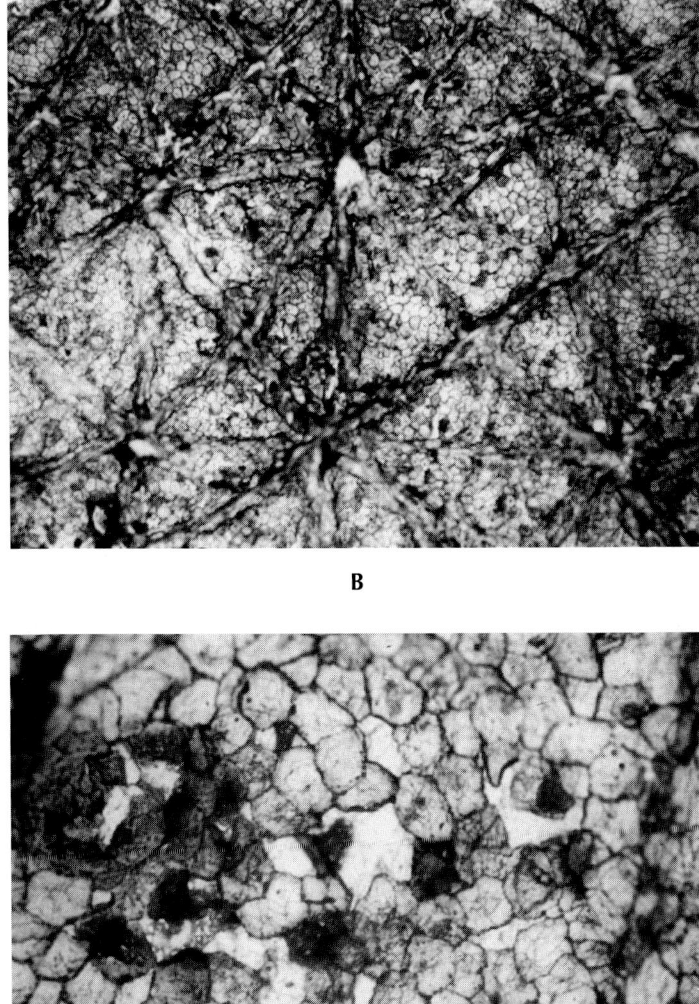

B

C

FIGURE 4 (continued)

Results may be illustrated under the microscope (Figure 4) and are expressed by two parameters named the corneosurfametry index (CSMI) and the colorimetric index of mildness (CIM).[47] CSMI is the color differential (d) between each test material and controls sprayed with water only. It is calculated by the following equation:

$$CSMI = [(dL^*)^2 + (dC^*)^2]^{1/2}$$

There is a linear negative correlation between values of L^* and Chroma C^*.[45] Accordingly, CIM of each sample is calculated following

$$CIM = L^* - C^*$$

In such evaluations, mild products have a low CSMI value and a high CIM value. The converse is true for harsh skin care or household products (Figure 4).

Corneosurfametry is also well suited to evaluate the efficacy of skin barrier creams. For that purpose, a standardized amount of the skin barrier product is applied onto the CSSS before spraying a reference surfactant such as SLS 1%.

C. CORNEOSURFAMETRY AND ATOPIC SENSITIVE SKIN

The term sensitive skin covers a broad spectrum of different conditions having in common the occurrence of adverse reactions or unwanted changes in response to external factors of the environment and/or related to personal care products. Some patients with atopic dermatitis are among the individuals with sensitive skin because they are more readily deranged by chemical and physical irritation. This is related to a defective barrier function.[48]

Corneosurfametry performed with a reference surfactant (SLS 1%) shows that CSSS from atopic patients often react more severely than those taken from normal individuals without sensitive skin (Figure 5). CIM is significantly lower in atopic patients (21 ± 6) than in normal nonsensitive individuals (29 ± 5), indicating a higher susceptibility to irritation induced by surfactants and some other skin care products in atopic patients.

D. POTENCY OF SQUAMOLYTIC AGENTS

CSSS, using a method similar to corneosurfametry, may be a substrate for the study of the bioactivity of squamolytic agents. The test product is applied on CSSS for various periods of time ranging from 15 min to 4 h. After abundant rinsing and staining with PMS, colorimetric values are recorded and CSMI and/or CIM parameters are calculated in a similar way as for corneosurfametry. The typical kinetics with α-hydroxyacids shows first an aggravation in the test values followed by a drop in these values, this latter phase indicating a loss of corneocytes from the CSSS (Figure 6). Optical profilometry done on the same specimens shows the immediate smoothing effect of the products with reduction in the value of Rz followed by desquamation and increased Rz (Figure 6).

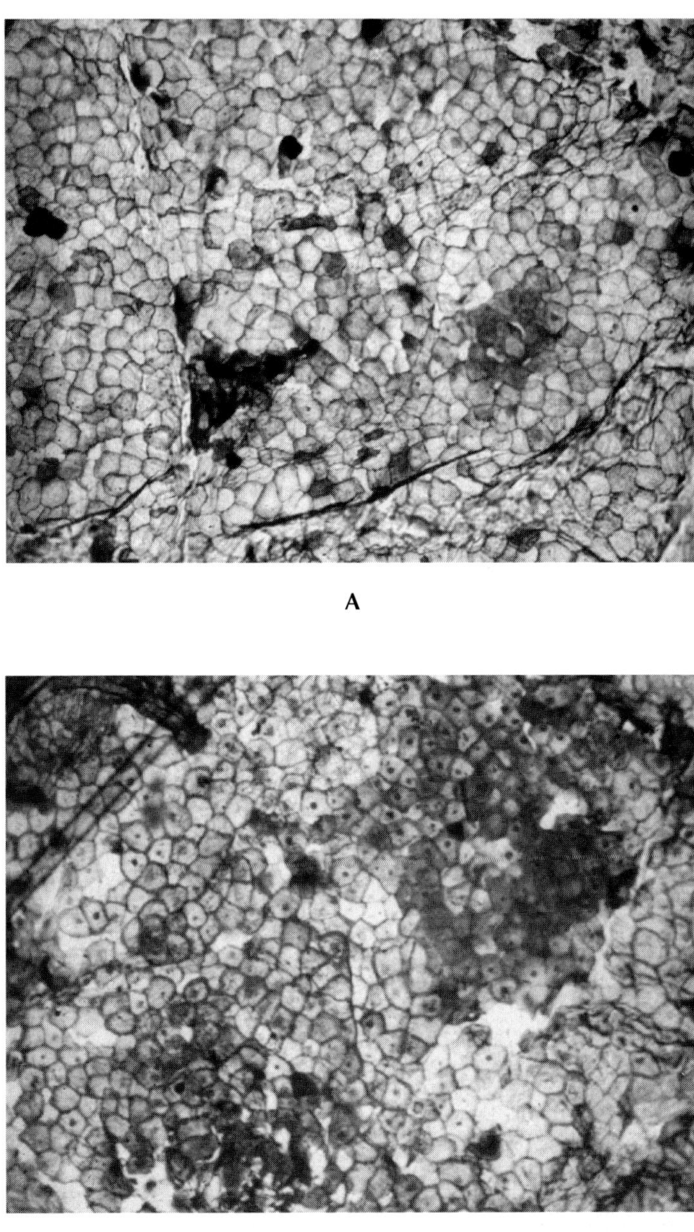

A

B

FIGURE 5 Microscopic aspect of a corneosurfametry test with 1% SLS on atopic skin. Both corneocytes (A) and parakeratotic cells (B) react strongly with the surfactant.

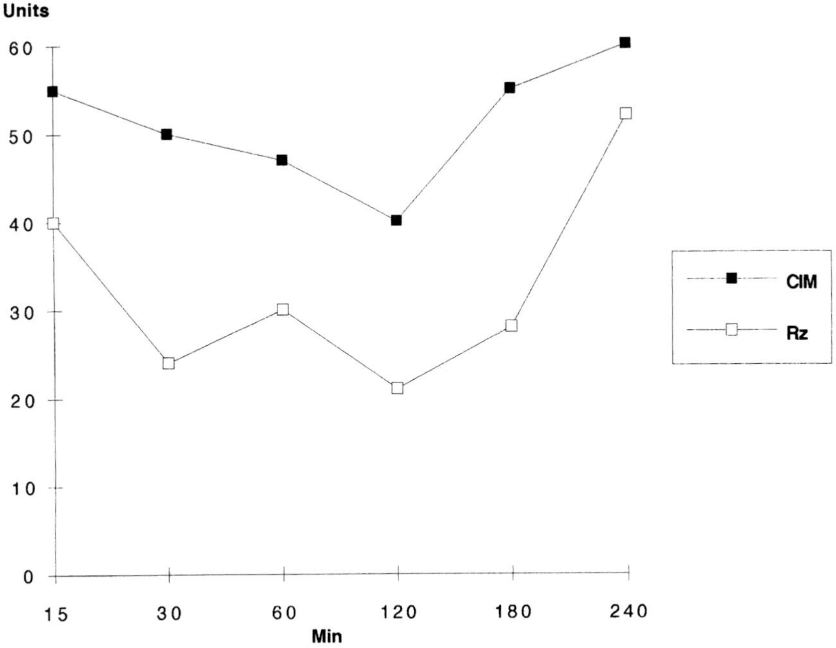

FIGURE 6   Kinetics of corneosurfametry (CIM) and skin roughness (Rz) of CSSS in contact with a 50% glycolic acid solution.

VIII. CONCLUSIONS

The possibilities of using skin strippings in pharmacology and cosmetology seem endless. The emergence of new analytical methods increases the validity of the information yielded by CSSS and SACD. With upcoming regulations avoiding animal experiments and ethical considerations for human testing, there is a need and a real possibility to develop new predictive bioassay using standardized skin strippings as substrate.

References

1. **Marks R. and Nicholls S.C.,** Drugs which influence the stratum corneum and techniques for their evaluation, *Clin. Exp. Dermatol.*, 6, 419, 1981.
2. **Goldschmidt, H. and Kligman, A.M.,** Exfoliative cytology of human horny layer. Methods of cell removal and microscopic techniques. *Arch Dermatol.*, 96, 575, 1967.

3. **Marks, R. and Dawber, R.P.R.,** Skin surface biopsy: an improved technique for the examination of the horny layer, *Br. J. Dermatol.*, 84, 117, 1971.

4. **Lachapelle, J.M., Gouverneur, J.C., Boulet, M., and Tennstedt, D.,** A modified technique (using polyester tape) of skin surface biopsy, *Br. J. Dermatol.*, 97, 49, 1977.

5. **Mills, O.H. and Kligman, A.M.,** The follicular biopsy, *Dermatologica*, 167, 57, 1983.

6. **Piérard-Franchimont, C. and Piérard, G.E.,** Assessment of aging and actinic damages by cyanoacrylate skin surface stripping, *Am J. Dermatopathol.*, 9, 500, 1987.

7. **Katz, H.I.,** Skin surface touch print: review of indications and uses, *Adv. Dermatol.*, 5, 197, 1990.

8. **Piérard, G.E., Piérard-Franchimont, C., and Dowlati, A.,** La biopsie de surface en dermatologie clinique et expérimentale, *Rev. Eur. Dermatol. MST*, 4, 455, 1992.

9. **Piérard, G.E., Piérard-Franchimont, C., Saint Léger, D., and Kligman, A.M.,** Squamometry: the assessment of xerosis by colorimetry of D-Squame adhesive discs, *J. Soc. Cosmet. Chem.*, 47, 297, 1992.

10. **Piérard-Franchimont, C. and Piérard, G.E.,** Skin surface stripping in diagnosing and monitoring inflammatory, xerotic and neoplastic diseases, *Ped. Dermatol.*, 2, 180, 1985

11. **Piérard-Franchimont, C. and Piérard, G.E.,** Les xéroses: structure de la peau rêche, *Int. J. Cosmet. Sci.*, 6, 47, 1984.

12. **Piérard, G.E.,** What do you mean by dry skin?, *Dermatologica*, 179, 1, 1989.

13. **Corcuff, P., de Rigal, J., Lévêque, J.L., Makki, S., and Agache, P.,** Skin relief and aging *J. Soc. Cosmet. Chem.*, 34, 177, 1983.

14. **Nikkels-Tassoudji, N., Piérard-Franchimont, C., De Doncker, P., and Piérard, G.E.,** Optical profilometry of nail dystrophies, *Dermatology*, 190, 301, 1995.

15. **Serup, J., Winther, A., and Blichmann, C.,** A simple method for the study of scale pattern and effects of a moisturizer. Qualitative and quantitative evaluation by D-Squame tape compared with parameters of epidermal hydration. *Clin. Exp. Dermatol.*, 14, 277, 1989.

16. **Schrader, K.,** Comparative experimental research on dandruff through quantitative image analysis, *J. Appl. Cosmetol.*, 4, 153, 1986.

17. **Kligman, A.M., Schatz H., Mannig, S., and Stoudemayer, T.,** Quantitative assessment of scaling in winter xerosis using image analysis of adhesive-coated disks (D-Squames), in *Noninvasive Methods for the Quantification of Skin Functions*, Frosch, P.J. and Kligman, A.M., Eds., Springer-Verlag, Berlin, 1993, 309.

18. **Piérard, G.E. and Piérard-Franchimont, C.,** Squamometry as an aid for rating severity of target lesions in atopic dermatitis, *Giorn. Int. Dermatol. Ped.*, 6, 125, 1994.

19. **Arrese, J.E., Piérard-Franchimont, C., De Doncker, P., Cauwenbergh, G., and Piérard, G.E.,** Effect of ketoconazole-medicated shampoos on squamometry and Malassezia ovalis load in pityriasis capitis, *Cutis,* in press.

20. **Holmes, R.L., Williams, M., and Cunliffe, W.J.,** Pilo-sebaceous duct obstruction and acne, *Br. J. Dermatol.*, 87, 327, 1972.

21. **Mills, O.H. and Kligman, A.M.,** A human model for assessing comedogenic substances, *Arch. Dermatol.*, 188, 903, 1982.

22. **Groh, D.G., Mills, O.H., and Kligman, A.M.,** Quantitative assessment of cyanoacrylate follicular biopsies by image analysis, *J. Soc. Cosmet. Chem.*, 43, 101, 1992.

23. **Piérard, G.E., Piérard-Franchimont, C., and Goffin, V.,** Digital image analysis of microcomedones, *Dermatology*, 190, 99, 1995.

24. **Sauermann, G., Ebens, B., and Hoppe, V.,** Analysis of facial comedos by porphyrin fluorescence and image analysis, *J. Toxicol. Cut.Ocular. Toxicol.*, 9, 369, 1990.

25. **Kligman, A.M.,** Regression method for assessing the efficacy of moisturizers, *Cosmet. Toilet*, 93, 27, 1978.

26. **Jansen, L.H., Hojko-Tomoko, M.T., and Kligman, A.M.,** Improved fluorescence staining technique for estimating turnover of the human stratum corneum, *Br. J. Dermatol.*, 90, 9, 1974.

27. **Piérard, G.E. and Piérard-Franchimont, C.,** Dihydroxyacetone test as a substitute for the dansyl chloride test, *Dermatology*, 186, 133, 1993.

28. **Piérard, G.E.,** Microscopic evaluation of the dansyl chloride test, *Dermatology*, 185, 37, 1992.

29. **Paye, M., Simion, A., and Piérard, G.E.,** Dansyl chloride labelling of stratum corneum: its rapid extraction from skin can predict irritation to surfactants and cleansing products, *Contact Derma.*, 30, 91, 1994.

30. **Lévêque, J.L., Porte, G., de Rigal, J., Corcuff, P., Francois, A.M., and Saint Léger, D.,** Influence of chronic sun exposure on some biophysical parameters of the human skin: an in vivo study, *J. Cut. Ag. Cosmet. Dermatol.*, 1, 123, 1988.

31. **Naka, W., Hanyaku, H., Tajima, S., Harada, T., and Nishikawa, T.,** Application of neutral red staining for evaluation of the viability of dermatophytes and Candida in human skin scales, *J. Med. Vet. Mycol.*, 32, 31, 1994.

32. **Piérard, G.E., Piérard-Franchmont, C., Van Cutsem, J., Rurangirwa, A., Hoppenbrouwers, M.L., and Schrooten, P.,** Ketoconazole 2% emulsion in the treatment of seborrheic dermatitis, *Int. J. Dermatol.*, 30, 806, 1991.

33. **Lücker, P.W., Beubler, E., Kukovetz, W.R., and Ritter, W.,** Retention time and concentration in human skin of bifonazole and clotrimazole, *Dermatologica*, 169 (Suppl. 1), 51, 1984.

34. **Gip, L.,** The in vitro determination of lingering antimycotic effect of two 1% omoconazole-nitrate cream formulations following single topical application, *Mycoses*, 31, 155, 1988.

35. **Pershing, L.K., Corlett, J., and Jorgensen, C.,** In vivo pharmacokinetics and pharmacodynamics of topical ketoconazole and miconazole in human stratum corneum, *Antimicrob. Agents Chemother.*, 38, 90, 1994.

36. **Knight, A.G.,** Culture of dermatophytes upon stratum corneum, *J. Invest. Dermatol.*, 59, 427, 1973.

37. **Aljabre, S.H.M., Richardson, M.D., Scott, E.M., and Schankland, G.S.,** Germination of Trichophyton mentagrophytes on human stratum corneum in vitro, *J. Med. Vet. Mycol.*, 30, 145, 1992.

38. **Rurangirwa, A. and Piérard, G.E.,** Culture de dermatophytes sur couche cornée, *Nouv. Dermatol.*, 8, 615, 1989.

39. **Rurangirwa, A., Piérard-Franchimont, C., and Piérard, G.E.,** Growth of Candida albicans on the stratum corneum of diabetic and non-diabetic patients, *Mycoses*, 33, 253, 1990.

40. **Rurangirwa, A., Piérard-Franchimont, C., and Piérard, G.E.,** Culture of fungi on cyanoacrylate skin surface strippings. A quantitative bioassay for evaluating antifungal drugs, *Clin. Exp. Dermatol.,* 14, 425, 1989.

41. **Piérard, G.E., Rurangirwa, A., and Piérard-Franchimont, C.,** Bioavailability of fluconazole and ketoconazole in human stratum corneum and oral mucosa, *Clin. Exp. Dermatol.,* 16, 168, 1991.

42. **Piérard, G.E., Piérard-Franchimont, C., and Arrese Estrada, J.,** Comparative study of the activity and lingering effect of topical antifungals, *Skin Pharmacol.,* 6, 208, 1993.

43. **Piérard, G.E., Arrese, J.E., and De Doncker, P.,** Antifungal activity of itraconazole and terbinafine in human stratum corneum. A comparative study, *J. Am. Acad. Dermatol.,* 32, 423, 1995.

44. **Arrese, J.E., Schrooten, P., De Doncker, P., Cauwenbergh, G., and Piérard, G.E.,** Fungal cultures on cyanoacrylate skin surface strippings as a dose-finding method for topical antifungals. A placebo-controlled study with itraconazole 0.25% and 0.50% creams, *J. Med. Vet. Mycol.,* 33, 127, 1995.

45. **Piérard, G.E., Goffin, V., and Piérard-Franchimont, C.,** Corneosurfametry: a predictive assessment of the interaction of personal-care cleansing products with human stratum corneum, *Dermatology,* 189, 152, 1994.

46. **Goffin, V., Paye, M., and Piérard, G.E.,** Comparison of in vitro predictive tests for irritation induced by anionic-surfactants, *Contact Dermat.,* in press.

47. **Piérard, G.E., Goffin, V., and Piérard-Franchimont, C.,** Squamometry and corneosurfametry for rating interactions of cleansing products with stratum corneum, *J. Soc. Cosmet. Chem.,* 45, 269, 1994.

48. **Tupker, R.A., Pinnagoda, J., Coenraads, P.J., Nater, J.P.,** Susceptibility to irritants: role of barrier function, skin dryness and history of atopic dermatitis, *Br. J. Dermatol.,* 123, 199, 1990.

11 Drug Concentration in the Skin

Christian Surber

TABLE OF CONTENTS

I. INTRODUCTION

An understanding of the pharmacodynamics and pharmacokinetics of a drug is essential for its safe use. A therapeutic effect of a drug and its appropriate dosage regimens are based upon the relations of time and drug concentration at the active site (target organ or biophase) and drug response. For reasons of feasibility, correlation of drug effects and drug concentration in blood or urine is most looked for. The three factors that are generally assessed are (1) the area under the serum concentration versus time curve, (2) the peak serum concentration, and (3) the time to peak serum concentration. Analytical difficulties sometimes preclude the measurement of drug or metabolite levels in the body fluids, in which case drug effects are assessed by observing an appropriate pharmacologic response.[1,2] Because blood and urine assays are available for nearly all drugs, and because pharmacokinetic methods are generally well understood, pharmacokinetics are the current mainstay of bioequivalence assessments. The key assumption is that equivalent pharmacokinetics indicate equivalent therapeutics. Or, more precisely, that observed differences between formulations for pharmacokinetic parameters are predictive of difference in clinical performance. However, little attempt to verify this key assumption is made.[3-5]

For topical drug products, e.g., ointments, creams, and gels (as distinguished from transdermal delivery systems), bioavailability studies used to establish bioequivalence of orally administered drugs are often difficult to carry out because drug concentration in blood may provide an inappropriate measure of bioequivalence. Furthermore topical doses tend to be so small that drug concentrations in blood and/or urine are often detectable using current assay techniques. Moreover, systemic availability may not properly reflect cutaneous bioavailability for medications intended to treat local skin orders.

Assessment of bioavailability or bioequivalence of topical formulations may be made by taking one of several possible approaches. Bioavailability or bioequivalence may be established through a well-controlled clinical trial or through a pharmacological or pharmacodynamic end point. Establishing that the concentration of drug within the treated tissue correlates with clinical performance or to some relevant pharmacological end point is another approach to bioavailability or bioequivalence that is currently being debated. So are some *in vitro* methods and animal studies that correlate with clinical performance.

At present a well-controlled clinical trial is the only uncontested, generally acceptable procedure for demonstrating bioavailability and bioequivalence of topical drug products. Because of the high costs, it is highly desirable to develop alternative *in vivo* (and *in vitro*) procedures. The purpose of this chapter is (1) to discuss some methods for direct drug concentration determination in the skin compartment, (2) to review the relationships between the drug concentration in the skin and a clinical, pharmacodynamic or pharmacological effect, and (3) to point out some difficulties in the interpretation of the results from these models.

II. SAMPLING TECHNIQUES

A. THE SUCTION BLISTER TECHNIQUE

The development of a defined suction blister located subepidermally was first described by Kiistala et al.[6-8] Apart from Kiistala's method to develop suction blisters using a special dome-shaped Dermovac® cap (Instrumentarium Corp., P.O. Box 67, Vitikka 1, SF-02630 Espoo, Finland) with several small holes (6 mm) suction blisters can also be developed using a plane block with holes of similar diameter (8 mm).[9] In a typical suction blister development with a Dermovac® a suction of about 200 mm Hg (2.66 Pa) below the atmospheric pressure is employed consistently. After a 2 to 3-h period 50- to 150-µl suction blister fluid and small stratum corneum–epidermis sheets can be harvested. The blister fluid corresponds roughly to the interstitial fluid. In contrast to the cantharidin blister fluid, another blister technique,[10] the suction blister fluid is free of white cells for the first 5 h; thereafter few polymophonuclear leukocytes appear.[11] The protein content of suction blisters is 60 to 70% of the corresponding serum value.[9] In diseased skin, blisters cannot be raised properly. Secretion of fluid into the capsule of the Dermovac® with and without bursting of the blister roof is observed. A standardization in diseased skin has not been established.

For pharmacokinetic studies up to 20 small blisters are formed. Drug is administered systemically or topically either before, during, or after blister raising. The blister fluid and blister roof are harvested at various time intervals for subsequent analysis.

Only erratic reports exist where the suction blister technique has been used to study topical drug delivery.[12-17] Treffel and co-workers[12] studied transepidermal absorption of citropten and bergapten from two different cosmetic tanning products (o/w emulsion, oil) into the suction blister fluid. The drug concentrations determined by high-pressure liquid chromatography were 37 and 51 ng/ml (w/o emulsion) and 26 and 23 ng/ml (oil), respectively, for citropten and bergapten.

A number of studies were published concerning 8-methoxypsoralen (8-MOP) absorption. Using the suction blister technique, Huuskonen et al.[13] studied absorption of 8-MOP during a PUVA bath therapy. More recently, Averbeck and co-workers[14] compared the photobiological activity[18] and the concentration of 5-MOP (bergapten) after topical application. They observed that the photobiological activity of 5-MOP in the suction blister fluid was more effective when it was delivered from an ethanolic solution than from a bergamot oil with the same 5-MOP content. From these experiments it can be concluded that the suction blister technique is useful not only for pharmacokinetic studies but also for pharmacodynamic evaluations, making it possible to study the biological activity of a drug and its metabolites at the level of the target organ.

Surber et al. and Laugier et al.[16,17] compared acitretin (retinoid) concentration in the skin after topical and systemic administration using the suction blister technique, and two other skin biopsy techniques. This data is discussed in Section II.J.

B.   THE CANTHARIDIN BLISTER TECHNIQUE

Large sheets of horny layer and blister fluid can be obtained with minor discomfort to the subjects using the cantharidin blister technique.[10] Many samples can be obtained from different regions of the same person with very little effort and with the advantage of not leaving a scar. Use is made of the capacity of cantharidin, a substance extracted from the Spanish Fly, to produce an intraepidermal blister. The cantharidin blister can be developed in man only. For most body areas a 0.2% solution of cantharidin in acetone is satisfactory. For sites where the stratum corneum is thick and dense the concentration is increased to 0.5%. The cantharidin acetone solution is placed into a glass or metal cylinder and the acetone evenly evaporated to dryness under a stream of air. A layer of wet cloth is placed over and beyond the application site and fastened occlusively to the skin by impervious plastic tape. Eight to ten h later, the turgid clear blister can be punctured using an insulin syringe to collect the blister fluid. Subsequently the blister itself can be cut away with scissors. Remnants of epidermis cling to the underface of the horny layer are removed by firm rubbing with a cotton-tipped applicator. Circular sheets of horny layer as large as 4 cm in diameter can be prepared. The amount of blister fluid is dependent on the size of the blister raised. Cantharidin blisters can also be obtained by the various cantharidin plasters described in old pharmacopoeias. The cantharidin blister fluid is an inflammatory exudate. It contains 650 to 12,700 white cells/μl.[19-21] Its albumin concentration significantly exceeds that of suction blister fluid (70 to 80% of the respective serum level).[22,21]

C.   RELATED TECHNIQUES

The surface layers of the skin are removed by suction blistering or by dermabrasion to obtain a skin window. In the case of suction blistering, the formed blister is excised. On the base of this skin window a chamber filled with saline solution or prewetted disks is applied. At various time intervals, the fluid or the disks are removed from the skin window and are refilled with isotonic saline or replaced by new disks.[23,24] The method of dermabrasion leads to an acute inflammatory reaction and large numbers of granulocytes and albumin are found in the skin chamber fluid. Access to other peripheral compartments is achieved by using different experimental "devices" that allow the collection of tissue fluid. Pertinent methods include the implantation of tissue cages, fibrin clots, and nylon or cotton threads.[25-29]

The blisters (suction blister and cantharidin blister) contain a fluid reservoir that communicates well with the intravascular fluid or the skin surface such that an immediate representation of the tissue concentration of a drug can be assumed. However, alterations of the composition of the blister fluid and the morphology of the blister base or blister roof may influence drug penetration into the blister fluid, e.g., the induction of an inflammatory response may result in a slower or enhanced passage of drug from the capillaries into the blister. The preservation of the barrier function of the stratum corneum following blister formation has also been questioned.[15] With respect to the morphological changes of the skin window the healing process also has to be taken into account.

D.   SURFACE RECOVERY

Percutaneous absorption may be determined by the loss of material from the surface as it penetrates into skin. Drug recovery from an ointment or a solution application is difficult because total recovery of compound from the skin can never be assured. Only in cases where a large amount of drug is absorbed the technique may be applicable. With the topical application of a transdermal delivery device, the entire unit can be removed from the skin and the residual amount of drug in the device can be determined. It is assumed that the difference between applied dose and residual dose is the amount of drug absorbed. It is also possible to monitor the disappearance of radioactivity from the surface of skin using an external Geiger–Müller counter.[30] The limitation of this methodology is that the disappearance is due both to the movement of ^{14}C-labeled chemical into the skin and to the quenching effect of the skin on the β-rays bouncing back to the instrument. The degree of quench of chemical in the various cell layers of the skin has not been defined. Lee and co-workers[31] recently developed a net methodology using ^{19}F-magnetic resonance spectroscopy (MRS) to measure the *in vivo* percutaneous absorption of flurbiprofen through hairless rat skin. A 2% w/v flurbiprofen gel containing isopropyl alcohol, water, and propylene glycol was used. Gel (100 mg) was applied to the skin of the lower back of an anesthetized hairless rat, contained with a rubber O-ring, and occluded with a plastic cover slip. The animal was placed on an MR surface coil and measurement were taken continuously over approximately 3 h in 10 min intervals with a 2-T GE CSI nuclear magnetic resonance (NMR) spectrometer. The disappearance of MR signal intensity per interval over time was measured. The measure was directly related to the percent of drug disappearance over time and was converted to a flux value.

Related to surface recovery technique, Phillips et al.[32-35] and Peck et al.[36] developed continuous transcutaneous drug collection (CTDC) device. The device consists of a salt-impregnated absorbent pad (e.g., 0.3 g agarose, 0.5 g activated carbon, 9.2 g 10% saline per 10 g[37]), which is covered by occlusive, water-impermeable tape. The method has been proposed as a method for studying drug disposition in subjects in a fashion that is less intrusive than currently available techniques, and which may be suited to the surveillance of drug exposure in ambulatory individuals. In particular, continuous transcutaneous drug collection has been proposed for the estimation of ethanol intake in drinking subjects[33] and for the monitoring of patient compliance with a drug regimen.[37]

Peck et al.[36] extensively analyzed the theoretical basis for the use of the CTCD device in assessing various aspects of drug pharmacokinetics. The time course of drug accumulation in the CTDC device was simulated using standard pharmacokinetic and mathematical techniques. The influence and implications of single and polyexponential drug disposition kinetics and backtransfer of drug from the collection device were investigated. Their interpretation of the results suggests two principal factors that affect the utility of this technique: First the degree of back transfer of drug from the CTDC device into the body. Second the duration of transcutaneous drug collection relative to the elimination rate of the drug from the body. In their theoretical treatment they excluded the notion that the rate of drug transfer into the CTDC device is somehow dependent upon the sweat secretion rate. They were aware of the fact that an occlusive dressing such as a CTDC device surely alters the skin permeability relative to unoccluded skin. However, they assume that the temperature and the hydration of the skin under the dressing become stabilized. Furthermore, their analyses suggested that the utility of a CTDC device is severely restricted when back transfer from the collection device is substantial. The back transfer is minimized or even excluded by introducing a "drug trap" binding the target chemical irreversibly; this would promote the use of this promising tool for assessing amount of drug exposure.

E.   SKIN SCRAPING

Large amounts of corneocytes can be harvested by curetting the stratum corneum from the surface with a surgical steel curette.[38] This technique has also been successfully employed by Faergemann et al. to determine drug concentration (terbinafine, fluconazole) in the stratum corneum.[39-41] The degree of superficial sebum and sweat contamination of the stratum corneum may influence the amount of drug found in the stratum corneum.

F.   SKIN SURFACE BIOPSY

Cyanoacrylate glue is applied to a glass slide and pressed onto the skin surface, and the slide is removed after the glue has polymerized.[42] This technique allows the quantification of the amount of drug found in the stratum corneum after various application times.[43] Because the hair follicle content also is removed with this method, a distinction between the transepidermal and the transfollicular routes of absorption is not possible.

G.   SKIN STRIPPING

The stripping method determines the concentration of chemical in the stratum corneum at the end of a short application period (30 min) and by linear extrapolation predicts the percutaneous absorption of that chemical for longer application periods. The chemical is applied to the skin of animals or humans and after the 30-min application time the stratum corneum is removed by successive tape application and removal. The tape strippings are assayed for chemical content. Dupuis, Lotte, Rougier, and co-workers[44-48] established a linear relationship between this stratum corneum reservoir content and *in vivo* percutaneous absorption using the standard urinary excretion method.[56-58] The major advantages of this method are the elimination of urinary and fecal excretion to determine absorption, and the applicability to nonradiolabeled determination of percutaneous absorption because the skin strippings contain adequate chemical concentrations for nonlabeled assay methodology. Despite the fact that the assay provides reliable prediction of total absorption, mechanistic interpretations are still rare. Based on the above data Auton[49] recently presented the first mathematical approaches that help to explain some of the above observations. Due to the potential of the tape-stripping technique (e.g., FDA Guidelines for Bioequivalence Testing of Topical Corticosteroids, 1992), a variety of parameters such as properties of adhesive tapes, vehicle effect on the harvesting of stratum corneum, and tape application and removal procedures are currently under systematic investigation.[50-53]

The technique has also been used to analyze pharmacodynamic activity, thus taking into account binding, decomposition, and metabolism of a given drug. Knight[54] stripped stratum corneum onto tapes, sterilized them, and then inoculated them with *Trichophyton mentagrohytes* spores. Under controlled conditions the spores germinate *in vitro* in the stratum corneum. Various formulations of topical griseofulvin were tested for activity and duration of effect with this method.

Sheth and co-workers[55] compared the concentration of iododeoxyurdine in the stratum corneum of the guinea pig skin by tape-stripping at different points after single and multiple topical doses of the drug in various *simple* formulations. These results were correlated with the efficacy of topical iododeoxyuridine against an experimental cutaneous herpes simplex virus. The results showed an excellent correlation between the quantity of iododeoxyuridine in the stratum corneum and the reduction in lesion severity ($r > 0.95$). Both formulation composition and dosing frequencies had an effect on the iododeoxyuridine concentration in the stratum corneum and the corresponding clinical efficacy.

Dupuis, Lotte, Rougier, and co-workers[45,48] could also show with a variety of *simple* pharmaceutical vehicles that the total amount of a drug (benzoic acid) absorbed (Feldmann/Maibach method[56-58]) could be predicted from the amount of the drug within the stratum corneum 30 min after application ($r = 0.93$).

More recently, Pershing and co-workers[52-60] simultaneously compared a skin blanching bioassay (Minolta CR-200 chromameter)[2] with drug content in human stratum corneum following topical application of four commercial 0.05% betamethasone dipropionate formulations (pharmaceutical *complex* vehicles). The rank order of the 0.05% betamethasone dipropionate formulation potency is similar between the visual skin blanching assay, the tape-stripping, and the *a* scale on the chromameter. The rank correlation between tape-stripping and the *a* scale on the chromameter was moderate to excellent.

H. SEBUM COLLECTION

Sebaceous glands are largest and most numerous on the forehead, face, in the ear, and on the midline of the back. Sebum is a complex mixture of lipids. Its principal components are glycerides, free fatty acids, wax esters, squalene, cholesterol, and cholesterol esters. Several functions have been ascribed to sebum, such as controlling water loss and protecting the skin from fungal and bacterial infection, however, these claims are also questioned. Sebum is miscible with water and allows water, polar, and nonpolar materials to penetrate. The length of time required for sebum to be produced and move from the base of the sebaceous gland to the skin surface is less than 10 days.[61-64]

Sebum is normally collected by swabbing the forehead, nose, checks, and midback with a sponge soaked in an organic solvent (i.e., isopropylalcohol, hexane). The sebum is extracted from the sponge for subsequent analysis. With other techniques a cigarette paper, a special tape, or a ground glass plate is pressed onto the skin site of interest to absorb the surface sebum.[40,65,66] Sebum collection has been extensively used to study the changes of the lipid composition during therapies with retinoids.

Attempts have also been made to detect drugs in the lipid mixture. Tetracycline and minocycline are well referenced and are clinically considered as effective drugs in the treatment of acne and other infectious dermatosis. Documentation of skin pharmacokinetic data is rare and controversial. Rashleigh et al.[67] showed that tetracycline detected by a fluorometric method accumulated in sebum after oral administration. Significant quantities were detected in the surface film taken from the forehead after 4 to 8 days. Tetracycline levels of 3 to 18 µg/g sebum lipid were detected. They did not measure tetracycline attained within the pilosebaceous lumens, in which acne centers. There was no direct relationship between the dose applied, blood levels, and concentration of the drug in the sebum. They state that from their experimental design and data a firm conclusion on the route of tetracycline to the skin cannot be drawn. With a microbiologic method Gould et al.[68] showed indirectly that in skin 70% of the terramycine was present compared with the serum terramycine level. Their method required approximately 1 g(!) skin from each subject. Aubin et al.[69] dosed 10 volunteers with minocycline (200 mg/day) during 4 weeks. They were not able to detect minocycline by high-pressure liquid chromatography (C_{18}) with a detection limit of 0.3 ng/ml. Luderschmidt et al.[70] successfully detected minocyline by high-pressure liquid chromatography (C_{18}) (no detection limit is given). In patients treated with 100 mg minocyline/day approximately 40 µg minocycline was detected on day 6 of the treatment on a 6 cm^2 area. With the exception of Rashleigh et al.[67] none of the investigators discusses the pathway of the drug to the skin. From the experiments carried out it is purely hypothetical that tetracyclines accumulate in the sebaceous glands.

The knowledge of the drug concentration in sebum is mainly derived from the examination of the surface samples collected from anatomical sites rich in sebaceous glands. An alteration of drug (and sebum) between the pilosebaceous lumen (and the site of biosynthesis) and the skin surface is highly probable since the corresponding time lag is about 8 to 9 days.[63,64] During that delay a number of factors could contribute to an alteration of initial sample such as (1) biotransformation by microorganisms or by enzymes released from the microorganisms; (2) dilution and/or reaction with species from another origin such as epidermal lipids; (3) absorption or binding to other media such as hair, or (4) alteration through environmental aggressions such as temperature, light, oxygen, and humidity. Isolation of sebaceous glands[71] from humans or animals during drug treatment is possible and may therefore be an alternative sebum collection procedure.

I.   HAIR AND NAIL COLLECTION

The use of hair or nails is advantageous in that it is easy to sample the tissue and xenobiotics remain unchanged in them for a long time period. And thus it is possible to monitor the intake of drugs and poisons or to follow the chronology of an intoxication. For forensic purposes the detection of arsenic in hairs is probably the most well known. As a result of better analytical equipment the detection of a variety of drugs in this matrix has been made possible for drugs such as phencyclidine, phenobarbital, opiates, amphetamines, tricyclic antidepressants, methadone, and nicotine.[72-77] These attempts have been mainly made in forensic medicine. In dermatology little effort has been made to detect drugs, e.g., antifungals, after systemic dosing in the hair.[78]

In a series of papers Walter and Flynn[79-83] developed a basic concept of the permeability of the human nail plate and thus created a better understanding of the toxic potentials and therapeutic possibilities of substances applied to the nail. The collective physiocochemical and clinical observations point to one general conclusion, that the nail is permeable for a variety of chemicals ranging from small polar and nonpolar nonelectrolytes and salts to large complex drug molecules. Collectively analyzed these studies indicate that the nail plate is intrinsically up to 1000 times more permeable to water than is stratum corneum, given that the ratio of thickness of these tissues is about 100 (nail to stratum corneum). Information from systematic investigations on penetration of various antifungals into the nail from topical formulations and the influence of the corresponding vehicles is still infrequent and incomplete.[84-86]

It is generally understood that in the planning of a drug regimen with, for example, an oral antibiotic for the treatment of an infection, it is necessary to know the pattern of serum and tissue levels following drug administration. When considered in parallel with information concerning the bactericidal levels of an antibiotic, it is possible to plan an appropriate oral dosage, application frequency, and duration of therapy. Although, in the past, there has been a lack of such relevant information concerning drugs used to treat disease of the nail plate, the same principles should apply. In a recent study with humans,[87] plasma terbinafine levels directly correlated with oral dose, and distal nail levels correlated with plasma levels. Terbinafine was generally detected earlier in proximal nail plate than in distal nail clippings, and the persistence of the drug in the nail plate paralleled plasma levels. Based on these data it should be possible to design a therapeutic regimen based on an optimal oral dose and duration of therapy.

J. SKIN BIOPSY

The most invasive—but still practical—method to access skin compartments is the excision of skin tissue. In contrast to the other methods described, the punch and shave biopsies allow a direct ingression into the compartment of interest. After removal (optional) of the stratum corneum from skin with an appropriate technique (tape-stripping, glue) the punch biopsy will represent parts of subcutaneous tissues, dermis, and epidermis and the shave biopsy will mainly represent epidermis and some dermis. Parts of the stratum corneum may remain on the epidermis depending on the method used for stratum corneum removal. Subcutaneous tissue can mechanically be divided from dermis and dermis can be separated from epidermis by heat.[88] Human skin samples larger than 100 mg are difficult to obtain and the usual amount that can be harvested is less than 50 mg.

Surber and Laugier[16,17] compared the acitretin concentration (synthetic retinoid) in human skin after oral and topical dosing using three different skin sampling techniques—punch biopsy, shave biopsy, and suction blister. All three skin sampling techniques have been used by various investigators to quantitate drugs and xenobiotics in the skin. Each technique gives access to distinctive skin compartments (Figure 1). The acitretin concentration in the skin after systemic application in a steady-state situation was comparable with the drug concentration reached after a single 24-h topical application of a saturated acitretin/isopropylmyristate formulation (Table 1). However, no beneficial effects in psoriasis and other disorders of keratinization are observed so far by the topical administration of acitretin. Similar observations have been made with methotrexate and ciclosporin.[89-90] One may postulate that drug concentration at a particular site within the skin following both routes of administration could be different due to the direction of the drug concentration gradient. This hypothesis has also been postulated and illustrated by Parry et al.[91]

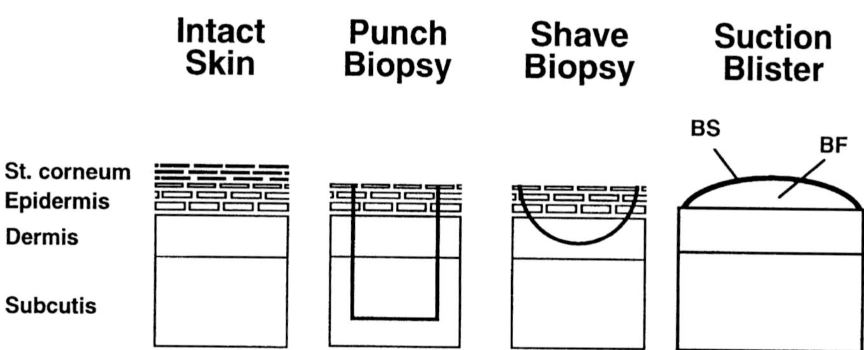

FIGURE 1 Schematic diagram of intact skin (left) and the skin specimens taken using three sampling techniques. BS; blister skin; BF, blister fluid. For details see text.

comparing the clinical efficacy of topical and oral aciclovir. Model predictions and *in vivo* data agree that topical administration of aciclovir results in a much greater total epidermal aciclovir concentration than after oral administration. However, mathematical modeling of aciclovir concentration gradient through the epidermis revealed that the drug concentration in the target site of the herpes simplex infection, the basal epidermis, was two to three times less after topical administration than after oral administration. Furthermore, one may postulate that drug metabolism could be different depending on the route of administration. Data supporting this hypothesis are still incomplete. Despite skilled experimentors, sophisticated sampling techniques, and instrumentations the information gained from these tissue samples is probably only an estimate of the chemical distribution within the skin. Because of possible interlaminate drug contamination, accurate and specific information on drug localization within a particular skin compartment following both routes of administration is not obtained by these methods.

The following two techniques have been used to obtain skin sections, parallel to the outer skin surface to a depth of about 300 to 600 μm.

TABLE 1

Total Drug Concentration in Skin after Oral and Topical Administration of Acitretin

	Total drug concentration (ng/g wet tissue)	
Sample	Oral application	Topical application
Punch biopsy	275	160
Shave biopsy	160	360
Suction blister fluid	200	90
Suction blister skin	460	3800

Data from Surber et al.[16] and Laugier et al.[17]

1. Semiautomated Skin Sectioning Technique

With this approach the skin tissue from a punch biopsy is placed onto a cryomicrotome table and 10- to 40-μm sections are cut parallel to the skin surface. This technique is excellently summarized in the form of a standard operating procedures by Schaefer et al.[92,93] When the cylinder-shaped punch biopsy is placed — dermis side down — on a microtome chuck maintained at $-17°C$ it is essential that the stratum corneum surface is parallel to the cutting plane. This can be accomplished when several metal rings of a total thickness slightly greater than that of the skin sample are placed around the tissue, and the rings are filled with embedding medium. Subsequently a flat, precooled surface (glass slide) is pressed against the stratum corneum, resulting in a flat outer surface. The whole assembly is then frozen in approximately 10 sec and can be stored. The technique does not take into account the dermal papillae and a clear separation of histologically distinct compartments is not possible. The method has predominately been used to characterize the pharmacokinetic and metabolic behavior of topically applied drugs.[94,95]

Using excised human skin and tissue grafted to athymic mice, the *in vitro* and *in vivo* concentration profiles of salicylic acid and salicylate esters were obtained for the outer several hundred microns of the skin.[95] The results show significant differences in the extent of enzymatic cleavage and distribution of metabolites between *in vitro* and *in vivo* studies. The data also suggest that *in vitro* results may overestimate metabolism because of increased enzymatic activity and/or decreased capillary removal (Figure 2).

2. Manual Skin Sectioning Technique

With the development of the skin sandwich flap model[96] Pershing and co-workers[97] proposed a new manual skin sectioning technique. With this technique the profile of disposition of test compounds in skin after topical administration is examined. Their technique requires the use of radiolabeled compound and fresh skin, without freezing. The sectioning technique utilizes cyanoacrylic cement, Scotch tape, glass slides, 115-μm-thick glass coverslips, a thin razor blade, and a fresh 2-mm skin punch biopsy. Briefly (Figure 3), a 2-mm skin punch biopsy is sectioned in 115-μm-thick sections starting from the stratum corneum (a) completely through the biopsy by dipping the stratum corneum end of the biopsy into cyanoacrylate adhesive (step 1) and placing the biopsy on a microscope slide between two 115-μm-thick microscope coverslips that are held in place on the microscope slide with tape (step 2). A single-edged razor blade is used to shave the stratum corneum side of the punch from the remainder of the biopsy. The remaining biopsy (b–d) is then redipped in the cyanoacrylate adhesive in the original orientation (stratum corneum face down) and placed in a new location on the microscope slide (b). The remainder of the

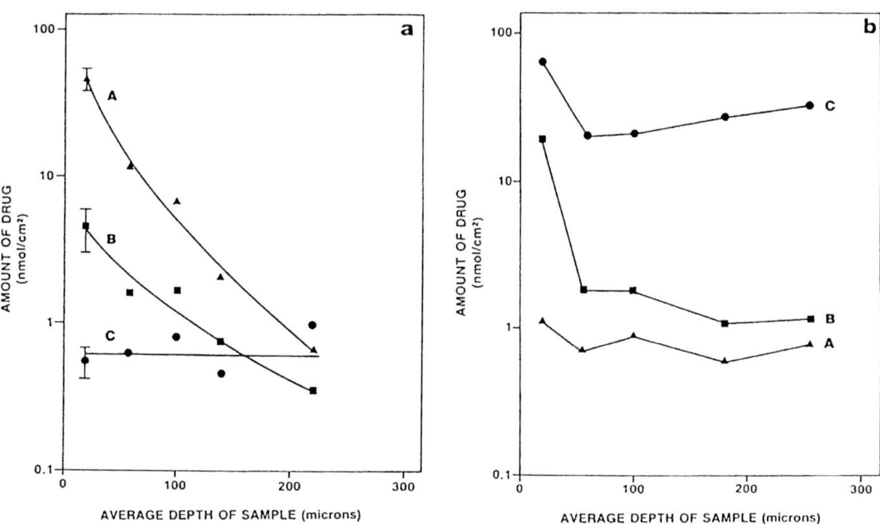

FIGURE 2   The concentration profile of the diester (A), monoester (B), and salicylate (C), as a function of the depth within human skin either grafted to athymic mice (a) or *in vitro* (b). Results were obtained after 24 h of topical drug application. (From Guzek et al.[95] with permission.)

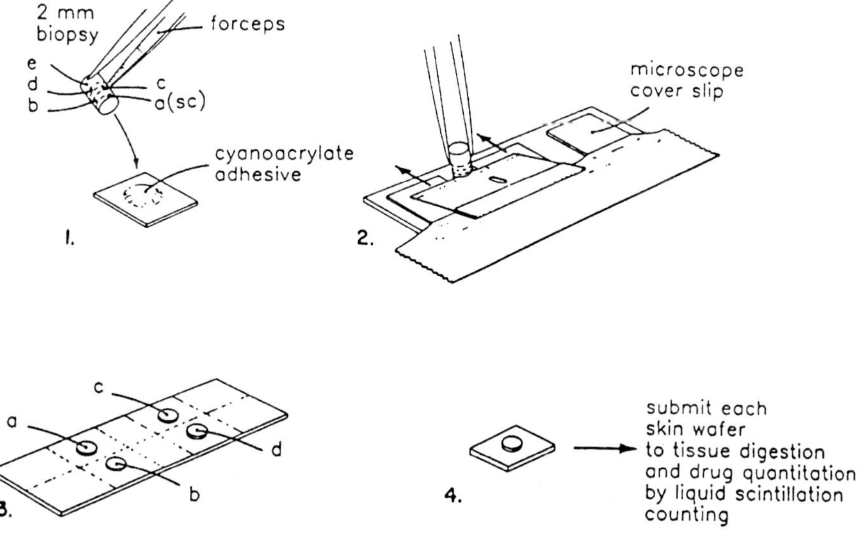

FIGURE 3   Manual sectioning technique of the punch biopsies. (From Pershing et al.[97] With permission.) For details see text.

skin biopsy is sectioned similarly, producing multiple 115-μm sections on an individual microscope slide (step 3). After complete sectioning of the skin biopsy, the various skin wafers on the microscope slide are individually digested with tissue soublizer and submitted to liquid scintillation counting for drug quantification and subsequently a concentration gradient and disposition profile of the test compound in the skin are reconstructed. The authors also report that freezing the 2-mm biopsy results in redistribution of the drug from the stratum corneum to the middle sections (400 to 600 μm depth) of the skin biopsy, thereby portraying an inaccurate distribution profile in the skin. Unfortunately this statement is not further documented. Eventhough the redistribution phenomenon was described only for the 2-mm biopsy it may also be relevant for the 4- to 6-mm biopsy. The statement make a careful reevaluation of the skin preparation procedure necessary.

K.   MICRODIALYSIS

Microdialysis is a sampling technique that allows continuous, real time drug monitoring and causes minimal tissue damage or physiological alterations to the test subject. Further advantages of microdialysis sampling include the fact that once a molecule crosses the dialysis membrane enzymatic degradation or protein binding is eliminated and the tissue can be directly sampled without causing significant tissue fluid loss. Additionally, thermal and hydrolytic degradation of the compounds of interest can be minimized by optimizing the microdialysis sampling rate and the nature of the dialysis medium.[98,99] A significant reduction in the number of animals required to perform dermal and transdermal drug delivery research may be possible. A single animal can be used for the continuous real-time study of drug concentration in the skin with microdialysis compared to numerous animals required for each time point by currently accepted methods.

The initial questions regarding the feasibility of this technique and the impact of the probe implantation on the flux of drug through the skin were addressed by studies using a commercially available probe implanted into full-thickness rat skin *in vitro*.[100] Results of the initial feasibility studies necessitated the development of a novel probe system based on minimal probe dimensions. While a coated probe of linear geometry has been used previously, a biocompatible silicone coating on a cellulose dialysis fiber for implantation into skin has been designed and validated by the same group.[97] Following topical application of 5-fluorouracil cream (Efudex®) the implanted dermal microdialysis probe successfully monitored the drug concentration in the skin *in vivo*. Edema, inflammation, and tissue scarring around the site of implantation assessed by histological examination appeared to be minimal during the first 24 to 36 h after implantation. The histological studies revealed lymphocyte infiltration into the area of implantation after 6 h and scar tissue formation around the probe after 72 h. While the problem of tissue damage due to probe implantation is a disadvantage of any invasive procedure, the extent of tissue damage determined by histological examination and drug perfusion studies was minimal.

In a very recent report Matsuyama et al.[101] successfully demonstrated that the intradermal implantation of a microdialysis probe *in vivo* was practical to study percutaneous absorption of methotrexate and to assess changes in drug concentrations following coapplication of an enhancer.

This simple method for determining drug concentration in the skin *in vivo* provides previously inaccessible information about drug transport in the skin and may serve in the future as a technique to study topical bioavailability and skin pharmacokinetics to guide drug development and patient therapy.

III. ANALYTICAL TECHNIQUES

A. AUTORADIOGRAPHY

Autoradiography is a photographic technique used to detect the localization of radioactive materials in specimens. The technique itself is over 100 years old, but its application to skin research has increased greatly during the past 30 years. The methodology has improved rapidly and application of the technique has progressed not only from a tissue level to cell level, but also to a subcellular level as well.[102] In the process of drug delivery to the skin, the localization of the penetrating/permeating compounds in the skin layers and the identification of transport pathways are essential. Only a few approaches to quantify the compound in the different skin layers and appendages have been reported. An important contribution to the quantitative evaluation was made by Schaefer and Schalla,[93,103] which introduced the skin section method (see above). To overcome certain disadvantages of their technique they supplemented their findings by qualitative histoautoradiography. Thus the combination of histoautoradiography and skin sections may point toward privileged permeation routes.

In a recent investigation[104] the development of an intracutaneous depot for drugs was studied using a new pharmaceutical ingredient (Transcutol® → diethylene glycol monoethyl ether). A *qualitative* assessment revealed that the distribution pattern of hydrocortisone (model drug studied) was dependent of the vehicle used. In the control group (treated with hydrocortisone without Transcutol®) the distribution was not uniform, whereas in the treatment group (treated with hydrocortisone with Transcutol®), the hydrocortisone was uniformly distributed in the skin and the amount of radioactivity present in the skin was higher in the treatment group relative to the control group.

In an autoradiographic study on percutaneous absorption of five ^{14}C-labeled oils (*n*-octadecane, decanoxy decane, 2-hexyldecanoxy octane, isopropyl myristate, and glyceryl trioleate) in the guinea pig a decreasing absorbability (the word "absorbability" was introduced by the author) in the following order was found: isopropyl myristate, glyceryl trioleate, *n*-octadecane, decanoxy decane, 2-hexyldecanoxy octane.[105] The skin irritation potentials (macroscopic observation of erythema) of these oils were in agreement with the absorbabilities observed by microautoradiography. Isopropyl myristate revealed much more severe irritation than 2-hexyldecanoxy octane.

Estradiol retention in the skin was investigated by Bidmon et al.[106] using dry mount autoradiography. ^{3}H-estradiol-17β in dimethyl sulfoxide, ethylene glycol, or sesame oil was applied to shaved rat skin in the dorsal neck region. The results of the experiments demonstrate that dry mount autoradiography provides both regional and cellular resolution for specific tissues and the penetration routes of diffusible compound. Since embedding and melting of tissue are avoided and permanent contact with photographic emulsion is obtained, translocation of labeled compound during the preparation of the autoradiograms is excluded and cellular and subcellular resolution is high. After 2 h of topical treatment with ^{3}H-estradiol-17β dissolved in dimethyl sulfoxide a distinct cellular distribution was apparent. Accumulation of radioactivity was found in epidermis, sebaceous glands, dermal papillae of hair, and fibroblasts. The stratum corneum accumulated and retained radioactivity, apparently forming a depot for the hormone. Strong concentration and retention of hormone were conspicuous in sebaceous glands for more than 24 h, suggesting that sebaceous glands serve as a second storage site for the hormone. In all autoradiograms two penetration pathways to the dermis were visible: one through the stratum corneum and epidermis, and the other through the hair canals and hair sheaths. The same permeation routes and sites of deposition and retention are recognizable for female and male rats independent of whether dimethyl sulfoxide, ethylene glycol, or sesame oil was used as a vehicle. It is important to note that low topical doses of drug reveal information on specific drug binding in target cells with cellular and subcellular resolution. Higher doses provide the advantage of visualizing permeation routes, gradients, and vehicle effects in the various skin compartments (Figure 4).

By means of a microcomputer-based image analysis autoradiographic method, it was possible to measure and visualize the effects of carriers and application time on the localization on the rat skin of two lipophilic compounds, tetrahydrocannabinol and oleic acid.[107] It was found that after 2 h, both compounds, dissolved in Transcutol®, show similar profile of penetration. No significant difference was found between the concentration in the epidermis and the appendages, either for tetrahydrocannabinol or oleic acid. After 24 h, a different behavior was described for each compound. Using (1) polyethylene glycol 400, (PEG$_{400}$, Merck), (2) diethylene glycol monoethyl ether, (Trancutol®, Gattefosse), and (3) propylene glycol (BHD)/ethanol (Frutarom) (7/3), a dramatic effect of formulation composition on the localization of tetrahydrocannabinol was observed (Figure 5).

These data document a first effort of quantitative autoradiography and the information conforms well with previously reported results of skin permeation behavior of tetrahydrocannabinol.[108]

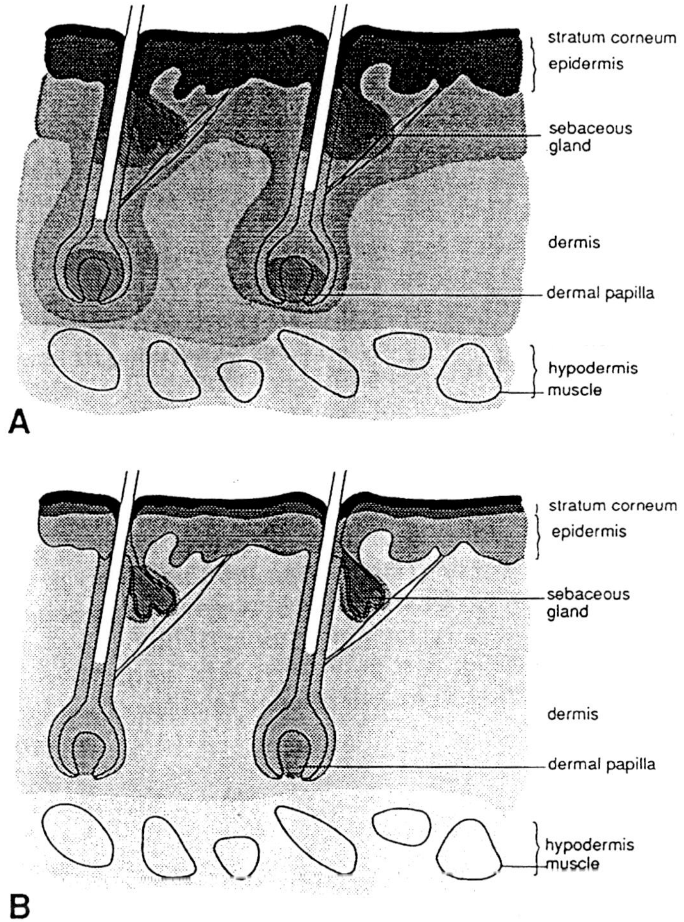

FIGURE 4 Scheme of distribution of radiolabeled estradiol 2 h (A) and 24 h (B) after topical application of 301 pmol/cm^2. The penetration gradients through the epidermis and the hair follicles are visible, as well as the enhanced uptake of radioactivity in sebaceous glands and dermal papillae. The highest amount of radioactivity is on top of the stratum corneum and in the stratum corneum (B). The different gray levels reflect the different silver grain densities. (From Bidman et. al.[106] With permission.)

Autoradiographic approaches are also used to study the effect of topically applied compounds (e.g., retinoic acid, Azone®, tetradecane) on epidermal growth using [^{3}H]tymidine.[109-111] In a recent study[112] autoradiographic and immunohistochemical techniques were combined to analyze the minoxidil localization in cultured vibrissa follicles revealing incongruent findings. The data suggest that minoxidil is not covalently bound to a cellular receptor in the hair follicle. Therefore, minoxidil's site of activity for stimulation of the hair follicle remains to be elucidated.

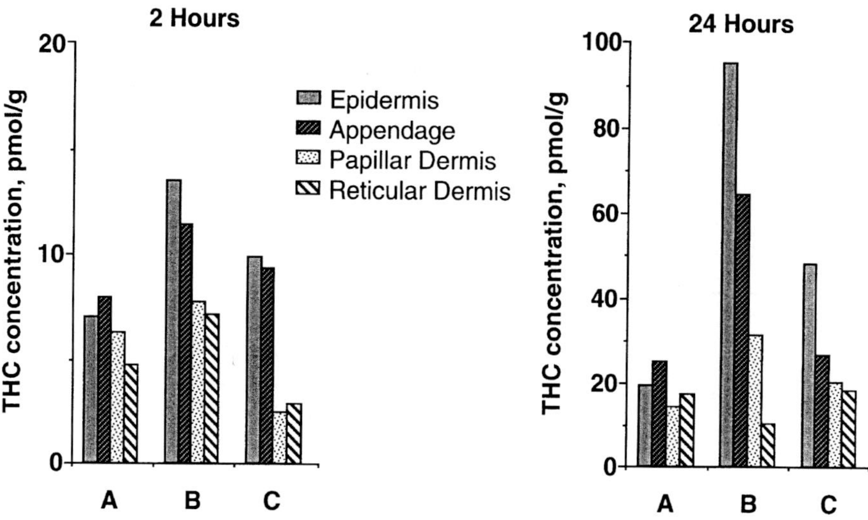

FIGURE 5 *In vivo* effect of vehicle and vehicle composition on the localization of tetrahydrocannabinol (THC) in the epidermis, appendages, and dermis of the hairless rat skin after 2 h and 24 h application. Vehicles: (A) polyethylene glycol 400 (PEG$_{400}$, Merck), (B) diethylene glycol monoethyl ether (Trancutol®, Gattefosse), (C) propylene glycol (BHD):ethanol (Frutarom) (7:3). (Redrawn from Fabin and Touitou.[107])

B. FLUORESCENCE

Fluorescent or fluorescence-labeled substances permit the use of the fluorescence microscope. In 1967 Borelli et al.[113] studied the *in vitro* percutaneous absorption of the fluorescence dye "acridinorange" incorporated into various commonly used topical preparations using domestic pig and human skin. Using fluorescence microscopy they observed qualitative different penetration characteristics of the dye into the skin depending on the preparations used. The addition of a nicotinic acid ester (not specified!) enhanced penetration of the dye into deeper skin structures. Follicles contributed significantly to the penetration of the dye. Meyer and co-workers[114] observed percutaneous absorption of tetracycline derivatives incorporated in various vehicles. Using semisolid formulations, penetration into deeper regions of the skin was minimal. With fluid formulations, penetration into deeper regions of the skin was minimal. With fluid formulations containing enhancers (dodecan, tetradecan) the penetration into skin was distinct. Fricker et al.[115] investigated the absorption of an intact oligopeptide in rat and dog small intestine using a stable somatostatin analogue. The octapetide, also a potential candidate for iontophoretic topical delivery, was coupled to 4-nitrobenzo-2-oxa-1,3-diazol to provide a fluorescent label for direct visualization. The labeled peptide was successfully used to investigate the existence of favored absorption sites in the small intestine. Although the biological qualities of the labeled and nonlabeled peptide were comparable, the derivatization of a molecule does change its physiocochemical qualities. Therefore such molecules are suitable to study only selected aspects of drug transport through skin. Fluorescent or fluorescence-labeled substances permit only the qualitative examination of permeation and penetration.

C. SAMPLE COLLECTION AND DRUG ANALYSIS: CRITICAL PARAMETERS

With a few exceptions tissue collection does involve a high degree of invasiveness and inconvenience and is therefore often impractical. Furthermore ethical considerations[116] in the conduct of human studies render it necessary to carefully select site of sampling, sample size, and frequency of sampling. Hence analytical assays and drug extraction procedure have to meet high demands.

Most of the above mentioned sampling techniques yield samples that reflect an instantaneous state of the drug in the tissue. However, changes of drug concentration in a specific compartment are necessary for a pharmacokinetic characterization of a drug. Analytical assays should be done according to standardized guidelines, e.g., "HPLC assays for bioavailability, bioequivalence and pharmacokinetic studies," a consensus of a symposium held in Washington D.C., December 3–5, 1990.[117] There is an important and unsolved problem with the determination of the drug extraction recovery from skin. To determine the extraction recovery from the skin tissue an *in vitro* experiment is usually performed where excised skin tissue is spiked with the drug. However the *in vivo* drug distribution into skin is different from the *in vitro* process and the information is probably only of limited value. The homogenization of certain tissues has been recognized as a major problem in drug analysis. Several methods are available to break up tissue (potter, dissolution with strong acids or bases, or enzymatic degradation). Skin tissue is one of the most difficult tissue structures to homogenize. We have evaluated several procedures and found that a homogenizer that combines mechanical and shock cavitation forces successfully breaks up the skin tissue structures without adding other problems to the sample treatment procedure, e.g., isomerization of metabolizing of the compounds of interest.[17]

D. C^* CONCEPT

A novel method based on the measurement of free drug concentration (C^*), or thermodynamic activity of drug, at the skin target site has been developed by Imanidis, Higuchi, Lee, and co-workers[118,119] to assess bioavailability and to predict efficacy for antivirals and other dermatological formulations. To measure C^* following the administration of a topical formulation entailed in a first step the establishment of a correlationship between the steady-state dermal drug flux and an elicited efficacy. This was accomplished by a novel animal model in which hairless mice were infected in a small spot at a lumbar skin area with cutaneous herpes simplex virus type 1. This induced, 3 days postinoculation, a narrow band of skin lesion development along the peripheral neural path toward the spinal cord. Taking advantage of this unique pattern of lesion development, an antiviral agent, such as aciclovir, was applied to an Azone®-pretreated skin area dorsal to the virus inoculation site and in the predicted path of lesion development to curtail more or less the lesion development. Five days after virus inoculation, the lesion development was scored for each mouse and two different antiviral efficacy were separately assessed: (1) "topical" (local) efficacy measured the antiviral activity to aciclovir delivery topically to the local skin area directly under the drug application site; and (2) "systemic" efficacy measured the antiviral activity of aciclovir delivery via systemic circulation to the target site, presumably the epidermal basal layer.[120] To quantify drug flux, a transdermal delivery system was developed in this animal model and the amount of aciclovir delivered to each infected animal could be controlled during the time period of drug therapy through a rate-controlling membrane. The actual (experimental) flux was determined at the end of each *in vivo* experiment by carrying out an extraction of the residual aciclovir in the transdermal delivery system. This extraction assay served to validate the expected (theoretical) flux or, alternatively, provided the bounds of uncertainty to the drug flux in the particular experiment. The results clearly showed a quantitative relationship between the antiviral efficacy and the experimental flux of aciclovir obtained from *in vivo* experiments. Topical efficacy increased with increasing aciclovir flux in the range of 10 to 100 $\mu g/cm^2$-day. Based on the relatively high precision of topical efficacy results, it is believed that the quantitative nature of this animal model should be valuable in the screening of new antiviral agents for topical treatment of cutaneous herpes virus infections and the optimization of topical formulations. Two factors may limit the applicability of this elegant approach to other classes of dermatological formulations. First, the drug was delivered from a transdermal delivery system at a constant rate over several days. With semisolid formulations, formulation application and formulation changes are difficult to control. And second, in the above approach the target for aciclovir (cutaneous herpes simplex virus type 1) and the localization of the target (epidermal basal layer) are known. Unfortunately the target sites for many dermatological agents are still unknown.

IV. CONCLUSION

In terms of the availability of topical drugs, the measurement of primary interest is the concentration of therapeutic agent within the skin or within a specific tissue layer of the skin. While some of the above techniques show considerable promise and recent evidence suggests that drug concentration in the skin correlates with measures of drug effect in skin, the techniques still require substantial further development and validation.[52,121] For this reason, current regulatory bioequivalence judgments have not been based on dermatopharmacokinetic studies. While it might be argued that acceptance of the method should require documentation of a good correlation to an observed clinical effect, this documentation has never been required for oral dosage forms to assess bioequivalence.

References

1. **Smolen, V.F., Williams, E.J., and Kuehn, P.B.,** Bioavailability and pharmacokinetic analysis of chlorpromazine in humans and animals using pharmacological data, *Can. J. Pharm.*, 10, 95, 1975.
2. **Queille-Roussel, C., Poncet, M., and Schaefer, H.,** Quantification of skin-colour changes induced by topical corticosteroid preparations using the Minolta Chroma Meter, *Br. J. Dermatol.*, 124, 1991.
3. **Somberg, J.C.,** Bioequivalence or therapeutic equivalence, *J. Clin. Pharmacol.*, 26, 1, 1986.
4. **Dettelbach, H.R.,** A time to speak out on bioequivalence and therapeutic equivalence, *J. Clin. Pharmacol.*, 25, 307, 1986.
5. **Lamy, P.P.,** Generic equivalents: issues and concerns, *J. Clin. Pharmacol.*, 26, 309, 1986.
6. **Kiistala, U.,** Suction blister device for separation of viable epidermis from dermis, *J. Invest. Dermatol.*, 50, 129, 1968.
7. **Kiistala, U.,** Dermal-epidermal separation. II. External factors in suction blister formation with special reference to the effect of temperature, *Ann. Clin. Res.*, 4, 236, 1972.
8. **Kiistala, U.,** Dermal-epidermal separation. I. The influence of age, sex and body region on suction blister formation in human skin, *Ann. Clin. Res.*, 4, 10, 1972.
9. **Schreiner, A., Hellum, K.B., Digranes, A., and Bergman, I.,** Transfer of penicillin G and ampicillin to human skin blisters induced by suction, *Scand. J. Infect. Dis.*, Suppl. 14, 233, 1978.
10. **Miescher, G.,** Beiträge zur Ekzemfrage. I. Zur Frage der Spezifität der ekzematösen Hautreaktion, *Arch. Dermatol. Syph.*, 173, 119, 1936.
11. **Kiistala, U., Mustakallio, K.K., and Rorsman, H.,** Suction blister in the study of cellular dynamics of inflammation, *Acta Dermatol. Venereol. (Stockholm)*, 47, 150, 1967.
12. **Treffel, P., Makki, S., Faivre, B., Humbert, P., Blanc, D., and Agache, P.,** Citropten and bergapten suction blister fluid concentrations after solar product application in man, *Skin Pharmacol.*, 4, 100, 1991.

13. **Huuskonen, H., Koulu, L., and Wilen, G.,** Quantitative determination of methoxalen in human serum, suction blister fluid and epidermis by gas chromatography mass spectrometry, *Photodermatology*, 1, 137, 1984.

14. **Averbeck, D., Averbeck, S., Blais, J., Moysan, A., Huppe, G., Morliere, B., Prognon, P., Vigny, P., and Dubertret, L.,** Suction blister fluid: its use for pharmacodynamic and toxicological studies of drugs and metabolites in vivo in human skin after topic or systemic administration, in *Models in Dermatology*, Maibach, H.I. and Lowe, N.J., Eds., S. Karger AG, Basel, 1989, 5.

15. **Agren, M.S.,** Percutaneous absorption of zinc from zinc oxide applied topically to intact skin in man, *Dermatologica,* 180, 36, 1990.

16. **Surber, C., Wilhelm, K.-P., Berman, D., and Maibach, H.I.,** In vivo skin penetration of acitretin in volunteers using three different sampling techniques, *Pharm. Res.*, 10, 1291, 1993.

17. **Laugier, J.-P., Surber, C., Bun, H., Geiger, J.-M., Durand, A., and Maibach, H.I.,** Determination of acitretin in the skin, in the suction blister, and in plasma of human volunteers after multiple oral dosing, *J. Pharm. Sci.*, 83, 623, 1994.

18. **Dubertret, L., Averbeck, D., Prognon, P., Blais, J., and Vigny, P.,** Photobiological activity of the suction blister fluid from patients treated with 8-methoxypsoralen, *Br. J. Dermatol.*, 109, 421, 1983.

19. **Allison, J.H. and Bettley, F.R.,** Investigations into cantharidin blisters raised on apparently normal skin in normal and abnormal subjects, *Br. J. Dermatol.*, 4, 330, 1958.

20. **Findlay, C.D., Wise, R., Allcock, J.E., and Durham, S.R.,** The tissue penetration as measured by a blister technique and pharmacokinetics of cefsulodin compared with carbenicillin and ticarcillin, *J. Antimicrob. Chemother.*, 7, 637, 1981.

21. **Wise, R., Gillett, A.P., Cadge, B., Durham, S.R., and Baker, S.,** The influence of protein binding upon tissue fluid levels of six beta-lactam antibiotics, *J. Infect. Dis.*, 142, 77, 1980.

22. **Simon, C., Malerczyk, V., Brahmstaedt, E., and Toeller, W.,** Cefazolin, ein neues Breitbandspektrum-Antibiotikum, *Dtsch. Med. Wochenschr.*, 98, 2448, 1973.

23. **Frongillo, R.F., Galuppo, L., and Moretti, A.,** Suction skin blister, skin window, and skin chamber techniques to determine extravascular passage of cefotaxime in humans, *Antimicrob. Agents Chemother.*, 19, 22, 1981.

24. **Shyu, W.C., Quintiliani, R., and Nightingale, C.H.,** An improved method to determine interstitial fluid pharmacokinetics, *J. Infect. Dis.*, 152, 1328, 1985.

25. **Raeburn, J.A.,** A review of experimental models for studying the tissue penetration of antibiotics in man, *Scand. J. Infect. Dis.*, Suppl. 14, 225, 1978.

26. **Ryan, D.M., Hodges, B., Spencer, G.R., and Harding, S.M.,** Simultaneous comparison of three methods for assessing ceftazidime penetration into extravascular fluid, *Antimicrob. Agents Chemother.*, 22, 995, 1982.

27. **Hoffstedt, B. and Walder, M.,** Penetration of ampicillin, doxycycline and gentamycin into interstitial fluid in rabbits and of penicillin V and pivampicillin in humans measured with subcutaneously implanted cotton threads, *Infection*, 9, 7, 1981.

28. **Holm, S.E.,** Experimental models for studies on transportation of antibiotics to extravasal compartments, *Scand J. Infect. Dis.*, 13, 47, 1978.

29. **Wise, R.,** Methods of evaluating the penetration of beta lactam antibiotics into tissues, *Rev. Infect. Dis.*, 8, 325, 1986.

30. **Marty, J.P., Bucks, D.A., and Maibach, H.I.,** Noninvasive radioisotope counting on skin: surface or external counting?, in *Percutaneous Absorption: Mechanisms—Methodology—Drug Delivery*, Bronaugh, R.L. and Maibach, H.I., Eds., Marcel Dekker, New York, 1989, 435.

31. **Lee, D.J., Burt, C.T., and Koch, R.L.,** Percutaneous absorption of flurbiprofen in the hairless rat measured in vivo using ^{19}F magnetic resonance spectroscopy, *J. Invest. Dermatol.*, 99, 431, 1992.

32. **Phillips, M.,** Sweat-patch test for alcohol consumption: rapid assay with an electrochemical detector, *Alcohol Clin. Exp. Res.*, 6, 532, 1982.

33. **Phillips, M. and McAloon, M.H.,** A sweat-patch test for alcohol consumption: evaluation in continuous and episodic drinkers, *Alcohol Clin. Exp. Res.*, 4, 391, 1980.

34. **Phillips, E.L.R., Little, R.E., Hillman, R.S., Labbe, R.F., and Campbell, C.,** A field test of the sweat patch, *Alcohol Clin. Exp. Res.*, 8, 233, 1984.

35. **Phillips, M.,** An improved adhesive patch for long-term collection of sweat, *Biomat. Med. Dev. Art. Org.*, 8, 13, 1980.

36. **Peck, C.C., Lee, K., and Becker, C.E.,** Continuous transepidermal drug collection: basis for use in assessing drug intake and pharmacokinetics, *J. Pharmacokinet. Biopharm.*, 9, 41, 1981.

37. **Peck, C.C., Conner, D.P., Bolden, B.J., Almirez, R.G., Kingsley, T.E., Mell, L.D., Murphy, G.M., Hill, V.E., Rowland, L.M., Ezra, D., Kwiatkowski, T.E., Bradley, C.R., and Abdel-Rahim, M.,** Outward transcutaneous chemical migration: implications for diagnostics and dosimetry, *Skin Pharmacol.*, 1, 14, 1988.

38. **Epstein, W. L., Shah, V.P., and Riegelman, S.,** Griseofulvin levels in stratum corneum. Study after oral administration in man, *Arch. Dermatol.*, 106, 344, 1972.

39. **Faergemann, J., Zehender, H., Denouel, J., and Millerioux, L.,** Levels of terbinafine in plasma, stratum corneum, dermis-epidermis (without stratum corneum), sebum, hair and nails during and after 250 mg terbinafine orally once per day for four weeks, *Acta Derm. Venereol. (Stockholm)*, 73, 305, 1993.

40. **Faergemann, J., Zehender, H., Jones, T., and Maibach, H.I.,** Terbinafine levels in serum, stratum corneum, dermis-epidermis (without stratum corneum, hair, sebum and eccrine sweat, *Acta Derm. Venereol. (Stockholm)*, 71, 322, 1991.

41. **Faergemann, J. and Laufen, H.,** Levels of fluconazole in serum, stratum corneum, epidermis-dermis (without stratum corneum) and eccrine sweat, *Clin. Exp. Dermatol.*, 18, 102, 1993.

42. **Marks, R. and Dawber, R.P.,** Skin surface biopsy: an improved technique for examination of the horny layer, *Br. J. Dermatol.*, 84, 117, 1971.

43. **Finlay, A. and Marks, R.,** Determination of corticosteroid concentration profiles in stratum corneum using the skin surface biopsy technique, *Br. J. Dermatol.*, 107, 33, 1982.

44. **Dupuis, D., Rougier, A., Roguet, R., Lotte, C., and Kalopissis, G.,** In vivo relationship between horny layer reservoir effect and percutaneous absorption in human and rat, *J. Invest. Dermatol.*, 82, 353, 1984.

45. **Dupuis, D., Rougier, A., Roguet, R., and Lotte, C.,** The measurement of the stratum corneum reservoir: a simple method to predict the influence of vehicles on in vivo percutaneous absorption, *Br. J. Dermatol.*, 115, 233, 1986.

46. **Rougier, A., Lotte, C., and Maibach, H.I.,** The hairless rat: a relevant animal model to predict in vivo percutaneous absorption in humans?, *J. Invest. Dermatol.*, 88, 577, 1987.

47. **Rougier, A., Lotte, C., and Dupuis, D.,** An original predictive method for in vivo percutaneous absorption studies, *J. Soc. Cosmet. Chem.*, 38, 397, 1987.

48. **Rougier, A., Rallis, M., Krien, P., and Lotte, C.,** In vivo percutaneous absorption: a key role for stratum corneum/vehicle partitioning, *Arch. Dermatol. Res.*, 282, 498, 1990.

49. **Auton, T.R.,** Skin stripping and science: a mechanistic interpretation using mathematical modelling of skin deposition as a predictor of total absorption, in *Prediction of Percutaneous Penetration*, Scott, R.C., Guy, R.H., Hadgraft, J., and Boddé, H.E., Eds., IBC Technical Services Ltd., London, 1990, 558.

50. **Tsai, J.C., Cappel, M.J., Weiner, N.D., Flynn, G.L., and Ferry, J.,** Solvent effects on the harvesting of stratum corneum for hairless mouse skin through adhesive tape stripping in vitro, *Intr. J. Pharm.*, 68, 127, 1991.

51. **Tsai, J., Weiner, N.D., Flynn, G.L., and Ferry, J.,** Properties of adhesive tapes used for stratum corneum stripping, *Int. J. Pharm.*, 72, 227, 1991.

52. **Pershing, L.K., Silver, B.S., Krueger, G.G., Shah, V.P., and Skelley, J.P.,** Feasibility of measuring the bioavailability of topical betamethasone dipropionate in commercial formulations using drug in skin and a skin blanching bioassay, *Pharm. Res.*, 9, 45, 1992.

53. **Henn, U., Surber, C., Schweitzer, A., and Bieli, E.,** D-Squame® adhesive tapes for standardized stratum corneum stripping, in *Prediction of Percutaneous Penetration*, Brain, K.R., James, V.J., and Walters, K.A., Eds., STS Publishing, London, 1993, 477.

54. **Knight, A.G.,** The activity of various topical griseofulvin preparations and the appearance of oral griseofulvin in the stratum corneum, *Br. J. Dermatol.*, 91, 49, 1974.

55. **Sheth, N.V., McKeough, M.B., and Spruance, S.L.,** Measurement of stratum corneum drug reservoir to predict the therapeutic efficacy of topical iododeoxyuridine for herpes simplex, *J. Invest. Dermatol.*, 89, 598, 1987.

56. **Feldmann, R.J. and Maibach, H.I.,** Percutaneous penetration of steroids in man, *J. Invest. Dermatol.*, 52, 89, 1969.

57. **Feldmann, R.J. and Maibach, H.I.,** Absorption of some organic compounds through the skin in man, *J. Invest. Dermatol.*, 54, 399, 1970.

58. **Feldmann, R.J. and Maibach, H.I.,** Percutaneous penetration of some pesticides and herbicides in man, *Toxicol. Appl. Pharmacol*, 28, 126, 1974.

59. **Pershing, L.K., Lambert, L.D., Shah, V.P., and Lam, S.Y.,** Variability and correlation of chromameter and tape-stripping methods with the visual skin blanching assay in quantitative assessment of topical 0.05% betamethasone dipropionate bioavailability in humans, *Int. J. Pharm*, 86, 201, 1992.

60. **Pershing, L.K., Corlett, J.L., Lambert, L.D., and Poncelet, C.E.,** Circadian activity of topical 0.05% betamethasone dipropionate in human skin in vivo, *J. Invest. Dermatol.*, 102, 734, 1994.

61. **Downing, D.T., Stewart, M.E., Wertz, P.W., Colton, S.W., Abraham, W., and Strauss, J.S.,** Skin lipids: an update, *J. Invest. Dermatol.*, 88, 2s, 1987.

62. **Rieger, M.D.,** Skin lipids and their importance to cosmetic science, *Cosmet. Toilet.*, 102, 36, 1987.

63. **Epstein, E.H. and Epstein, W.L.,** New cell formation in human sebaceous glands, *J. Invest. Dermatol.*, 46, 453, 1966.

64. **Downing, D.T., Strauss, J.S., Ramasastry, P., Abel, M., Lees, C.W., and Pochi, P.E.,** Measurement of the time between synthesis and the surface excretion of sebaceous lipids in sheep and man, *J. Invest. Dermatol.*, 64, 215, 1975.

65. **Nordstrom, K.M., Schmus, H.G., McGinley, K.J., and Leyden, J.J.,** Measurement of sebum output using a lipid absorbent tape, *J. Invest. Dermatol.*, 87, 260, 1986.

66. **Saint-Leger, D. and Bague, A.,** A simple and accurate routine procedure for qualitative analysis of skin surface lipids in man, *Arch. Dermatol. Res.*, 271, 215, 1981.

67. **Rashleigh, P.L., Rife, E., and Goltz, R.W.,** Tetracycline levels in skin surface film after oral administration of tetracycline to normal adults and to patients with acne vulgaris, *J. Invest. Dermatol.*, 49, 611, 1967.

68. **Gould, J.C. and Richtie, H.D.,** The terramycin content of human skin during therapy: a comparison with serum and urine levels, *Br. J. Plast. Surg.*, 5, 208, 1952.

69. **Aubin, F., Blanc, D., Guinchard, C., and Agache, P.,** Absence de passage de la minocycline dans le sebum?, *Ann. Dermatol. Venereol.*, 115, 977, 1988.

70. **Luderschmidt, C., Nissen, H., Neubert, U., and Knuechel, M.,** Newer methods for measuring therapeutic response in acne, *Front. Dermatol.*, 12, 131, 1983.

71. **Kellum, R.E.,** Isolation of human sebaceous glands, *Arch. Dermatol.*, 93, 610, 1966.

72. **Suzuki, O., Hattori, H., and Asano, M.,** Nails as useful materials for detection of methamphetamine or amphetamine abuse, *Forensic Sci. Int.*, 24, 9, 1984.

73. **Ochsendorf, F.R., Runne, U., Schöfer, H., Schmidt, K., and Raudonat, H.W.,** Sequential chloroquine concentration in human hair represent ingested dose and duration of therapy—human hair as a pharmacologic and toxicologic tachograph, *Arch. Dermatol.*, 281, 121, 1989.

74. **Balabanova, S. and Wolf, H.U.,** Determination of methadone in human hair by radioimmunoassay, *Z. Rechtsmed.*, 102, 1, 1989.

75. **Ishiyama, I., Nagai, T., and Toshida, S.,** Detection of basic drugs (methamphetamine, antidepressants and nicotine) from human hair, *J. Forensic Sci.*, 28, 380, 1983.

76. **Smith, F.P. and Liu, R.H.,** Detection of cocaine metabolite in perspiration stain, menstrual bloodstain, and hair, *J. Forensic Sci.*, 31, 1269, 1986.

77. **Balabanova, S., Brunner, H., and Nowak, R.,** Radioimmunological determination of cocaine in human hair, *Z. Rechtsmed.*, 98, 229, 1987.

78. **Van Cutsem, J., Van der Flaes, M., Thienpont, D., Dony, J., and Hörig, C.,** Quantitative Bestimmung von Ketoconazol in den Haaren oral behandelter Ratten und Meerschweinchen, *Mykosen*, 23, 418, 1980.

79. **Walters, K.A., Flynn, G.L., and Marvel, J.R.,** Physiocochemical characterization of the human nail: permeation pattern for water and the homologous alcohols and differences with respect to the stratum corneum, *J. Pharm. Pharmacol.*, 35, 28, 1983.

80. **Walters, K.A., Flynn, G.L., and Marvel, J.R.,** Physiochemical characterization of the human nail: solvent effects on the permeation of homologous alcohols, *J. Pharm. Pharmacol.*, 37, 771, 1985.

81. **Walters, K.A., Flynn, G.L., and Marvel, J.R.,** Penetration of the human nail plate: the effects of the vehicle pH on the permeation of miconazole, *J. Pharm. Pharmacol.*, 37, 498, 1985.

82. **Walters, K.A.,** Permeability characteristics of the human nail plate, *Int. J. Cosmet. Sci.*, 5, 231, 1983.

83. **Walters, K.A., Flynn, G.L., and Marvel, J.R.,** Physiochemical characterization of the human nail. I. Pressure sealed apparatus for measuring nail plate permeability, *J. Invest. Dermatol.*, 76, 76, 1981.

84. **Stüttgen, G. and Bauer, E.,** Bioavailability, skin- and nail penetration of topically applied antimycotics, *Mykosen*, 25, 74, 1982.

85. **Pittrof, F., Gerhards, J., Erni, W., and Klecak, G.,** Loceryl® nail lacquer—realization of a new galenical approach to onychomycosis therapy, *Clin. Exp. Dermatol.*, 17, 26, 1992.

86. **Roncari, G., Ponelle, C., Zumbrunnen, R., Guenzi, A., Dingemanse, J., and Jonkman, J.H.,** Percutaneous absorption of amorolfine following a single topical application of an amorolfine cream formulation, *Clin. Exp. Dermatol.*, 17, 33, 1992.

87. **Dykes, P.J., Thomas, R., and Finlay A. Y.,** Determination of terbinafine in nail samples during systemic treatment for onychomycoses, *Br. J. Dermatol.*, 123, 481, 1990.

88. **Surber, C., Wilhelm, K.-P., Hori M., Maibach, H.I., and Guy, R.H.,** Opitimization of topical therapy: partitioning of drugs into stratum corneum, *Pharm. Res.*, 7, 1320, 1990.

89. **Fry, L. and McMinn, R.M.,** Topical methotrexate in psoriasis, *Arch. Dermatol*, 96, 483, 1967.

90. **Surber, C., Itin, P., and Büchner, S.,** Clinical controversy on the effect of topical ciclosporin A: what is the target organ?, *Dermatology*, 185, 242, 1992.

91. **Parry, G.E., Dunn, P., Shah, V.P., and Pershing, L.K.,** Acyclovir bioavailability in human skin, *J. Invest. Dermatol.*, 98, 856, 1992.

92. **Schaefer, H., and Lamaud, E.,** Standardization of experimental models, in *Skin Pharmacokinetics*, Shroot, B. and Schaefer, H., Eds., Karger, Basel, 1987, 77.

93. **Schaefer, H., Stüttgen, G., Zesch, A., Schalla, W., and Gazith, J.,** Quantitative determination of percutaneous absorption of radiolabeled drugs in vitro and in vivo by human skin, *Curr. Probl. Dermatol.*, 7, 80, 1978.

94. **Zesch, A. and Schäfer, H.,** Penetrationskinetik von radiomarkierten Hydrocortisone aus verschiedenartigen Salbengrundlagen in die menschliche Haut, *Arch. Derm. Forsch.*, 252, 245, 1975.

95. **Guzek, D.B., Kennedy, A.H., McNeill, S.C., Wakshull, E., and Potts, R.O.,** Transdermal drug transport and metabolism. I. Comparison of in vitro and in vivo results, *Pharm. Res.*, 6, 33, 1989.

96. **Wojciechowski, Z., Pershing, L.K., Huether, S., Leonard, L.B., Higuchi, W.I., and Krueger, G.G.,** An experimental skin sandwich flap on an independent vascular supply for the study of percutaneous absorption, *J. Invest. Dermatol.*, 88, 439, 1987.

97. **Pershing, L.K. and Krueger, G.G.,** Human skin sandwich flap model for percutaneous absorption, in *Percutaneous Absorption*, Bronaugh, R.L. and Maibach, H.I., Eds., Marcel Dekker, New York, 1989, 397.

98. **Scott, D.O., Bell, M.A., and Lunte, C.E.,** Microdialysis-perfusion sampling for the investigation of phenol metabolism, *J. Pharm. Biomed. Analysis*, 7, 1249, 1989.

99. **Lunte, C.E., Scott, D.O., and Kissinger, P.T.,** Sampling living systems using microdialysis, *Anal. Chem.,* 63, 773A, 1991.

100. **Ault, J.M., Lunte, C.E., Meltzer, N.M., and Riley, C.M.,** Microdialysis sampling for investigation of dermal drug transport, *Pharm. Res.,* 9, 1256, 1992.

101. **Matsuyama, K., Nakashima, M., Nakaboh, Y., Ichikawa, M., Yano, T., and Satoh, S.,** Application of in vivo microdialysis to transdermal absorption of methotrexate in rats, *Pharm. Res.,* 11, 684, 1994.

102. **Fukuyama, K.,** Autoradiography, in *Methods in Skin Research,* Skerrow, D. and Skerrow, C.J., Eds., John Wiley, London, 1985, 71.

103. **Schalla, W., Jamoulle, J., and Schaefer, H.,** Localization of compounds in different skin layers and its use as an indicator of percutaneous absorption, in *Percutaneous Absorption: Mechanism—Methodology—Drug Delivery,* Bronaugh, R.L. and Maibach, H.I., Eds., Marcel Dekker, New York, 1989, 283.

104. **Ritschel, W.A., Panchagnula, R., Stemmer, K., and Ashraf, M.,** Development of an intracutaneous depot for drugs, *Skin Pharmacol.,* 4, 235, 1991.

105. **Suzuki, M., Asaba, K., Komatsu, H., and Mochizuka, M.,** Autoradiographic study on percutaneous absorption of several oils useful for cosmetics, *J. Soc. Cosmet. Chem.,* 29, 265, 1978.

106. **Bidmon, H., Pitts, J.D., Solomon, H.F., Bondi, J.V., and Stumpf, W.E.,** Estradiol distribution and penetration in rat skin after topical application, studied by high resolution autoradiography, *Histochemistry,* 95, 43, 1990.

107. **Fabin, B. and Touitou, E.,** Localization of lipophilic molecules penetrating rat skin in vivo by quantitative autoradiography, *Int. J. Pharm.,* 74, 59, 1991.

108. **Touitou, E., Fabin, B., Dany, S., and Almog, S.,** Transdermal delivery of tetrahydrocannabinol, *Int. J. Pharm.,* 43, 9, 1988.

109. **Plewig, G. and Fulton, J.E.,** Autoradiographische Untersuchungen an Epidermis und Adnexen nach Vitamin A-Säure-Behandlung, *Hautarzt,* 23, 128, 1974.

110. **Klein, J.A., McCullough, J.L., and Weistein, G.D.,** Topical tritiated thymidine for epidermal growth fraction determination, *J. Invest. Dermatol.,* 86, 406, 1986.

111. **Motoyoshi, K., Ito, M., Sakamoto, F., and Sato, Y.,** A time course study of the proliferation of sebaceous glands induced by topically applied tetradecane in rabbit pinna skin: autoradiography and electron microscopy, *J. Dermatol.,* 14, 9, 1987.

112. **Zelei, B.V., Walker C.J., Sawada, G.A., Kawabe, T.T., Knight, K.A., Buhl, A.E., Johnson, G.A., and Diani, A.R.,** Immunohistochemical and autoradiographic findings suggest that minoxidil is not localized in specific cells of vibrissa, pelage, or scalp follicles, *Cell Tissue Res.,* 262, 407, 1990.

113. **Borelli, S.M.,** Fluorescenzmikroskopische Untersuchungen über die perkutane Penetration fluorescierender Stoffe, *Hautarzt,* 8, 261, 1957.

114. **Meyer, F.,** Örtliche Anwendung von Tetracyclinen, *Arch. Pharmakol. Exp. Pathol.,* 225, 47, 1966.

115. **Fricker, G., Bruns, C., Munzer, J., Briner, U., Albert, R., Kissel, T., and Vonderscher, J.,** Intestinal absorption of octapeptide SMS 201-995 visualized by fluorescence derivatization, *Gastroenterology,* 100, 1044, 1991.

116. **Svensson, C.K.,** Ethical considerations in the conduct of clinical pharmacokinetic studies, *Clin. Pharmacokinet.,* 17, 217, 1989.

117. **Shah, V.P., Midha, K.K., Dighe, S., McGilveray, I.J., Skelly, J.P., Yacobi, A., Layloff, T., Viswanathan, C.T., Cook, C.E., McDowall, R.D., Pittman, K.A., and Spector, S.,** Analytical methods validation: bioavailability, bioequivalence and pharmacokinetic studies, *Pharm. Res.*, 9, 588, 1992.

118. **Imanidis, G., Song, W., Lee, P.H., Su, M.H., Kern, E.R., and Higuchi, W.I.,** Estimation of skin target site acyclovir concentrations following controlled (trans)dermal drug delivery in topical and systemic treatment of cutaneous HSV-1 infections in hairless mice, *Pharm. Res.*, 11, 1035, 1994.

119. **Lee, P.H., Su, M.H., Kern, E.R., and Higuchi, W.I.,** Novel animal model for evaluating topical efficacy of antiviral agents: flux versus efficacy correlations in acyclovir treatment of cutaneous herpes simplex virus type 1 (HSV-1) infections in hairless mice, *Pharm. Res.*, 9, 979, 1992.

120. **Price, R.W.,** Neurobiology of human herpes virus infections, *Crit. Rev. Clin. Neurobiol.*, 2, 61, 1986.

121. **Caron, D., Queille-Roussel, C., Shah, V.P., and Schaefer, H.,** Correlation between the drug penetration and the blanching effect of topically applied hydrocortisone creams in human beings, *J. Am. Acad. Dermatol.*, 23, 458, 1990.

12 The Plastic Occlusion Stress Test (POST) as a Model to Investigate Skin Barrier Function

Enzo Berardesca and Howard I. Maibach

The epidermal permeability barrier is located in the intercellular spaces of the stratum corneum[1] and is organized into multilayers with lamellar repeat structure.[2] Evaluation and quantification of barrier function efficiency are important in investigating skin physiology, pathology, and toxipharmacology. Several techniques can be utilized to investigate barrier properties; *in vitro*, qualiquantitative analysis of intercellular lipids of the stratum corneum provides a reliable estimate of the permeability characteristics of a particular skin[3,4]; *in vivo*, stratum corneum integrity and barrier function can be measured indirectly by transepidermal water loss (TEWL) recordings.[5] Baseline TEWL measurements have also been extensively used to quantify irritant reactions and contact dermatitis.[6]

Stress tests provide additional information as compared to steady measurements and may be useful in investigative dermatology to study skin pathophysiology by increasing the dynamics of skin behavior and skin responses. Occlusion of the skin causes an immediate increase in relative skin moisture and therefore an increase of TEWL when measured after the removal of occlusion.[7,8] The plastic occlusion stress test is a method that utilizing skin occlusion allows the noninvasive quantification of some functional parameters of the stratum corneum efficiency.

This test consists in applying a plastic chamber on the skin surface and sealing it firmly to induce an occlusion to increase the water content underneath the chamber. Hydration is proportional to the occlusion time: prolonged occlusion (>3 days) induces a TEWL increase lasting several days after the removal of the occlusion, even though relative stratum corneum water content is decreased.[9] In our experience we have seen that 24 h is very practical for our work and gives good results in terms of water accumulation and discriminating

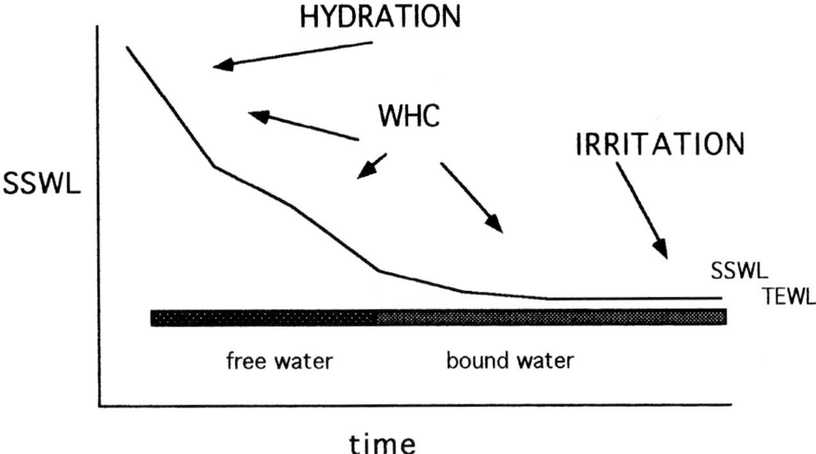

FIGURE 1 Influence of some functional properties of the stratum corneum on the shape of the POST curve (see text for details).

power of the decay curves obtained. When the occlusion is removed, the excess of water on the skin starts to evaporate and can be recorded using an evaporimeter. This evaporation can be defined as skin surface water loss (SSWL) rather than TEWL: indeed, it represents the evaporation of water trapped within and over the stratum corneum and not the water passing through the stratum corneum form the viable epidermis.

Information related to skin physiology and function can be obtained by analyzing plastic occlusion stress test (POST) decay curves, namely; stratum corneum hydration, integrity of the barrier function, water-holding capacity (WHC), and free and bound water compartments.

Figure 1 shows a typical dehydration curve obtained with the POST technique. The amount of water evaporating during the first minutes is directly proportional to skin hydration. The higher the stratum corneum water content, the higher is the degree of evaporation in the air. At the end of the dehydration time, SSWL is greatly reduced and the evaporation detected by the evaporimeter probe is mainly due to the TEWL. Therefore, changes induced in this part of the curve are due to damage of the water barrier and reflect the condition of the barrier function. The shape and the slope of the curve describe the total amount of water evaporated and, therefore, the WHC of the stratum corneum. The evaluation of the slope is rather difficult since the curve shows a biexponential pattern of decay and can be fitted using a biexponential regression analysis. However, an easier approach consists in transforming the data in log to have a more linear decay and then to use a simple regression analysis.

Free and bound water compartments of the stratum corneum can be estimated by analyzing POST decay curves (Figure 1). Indeed, the evaporation of free water is recorded during the first 10 min of the decay curve, where the curve is steeper (first exponential). Later, from the tenth minute onward the

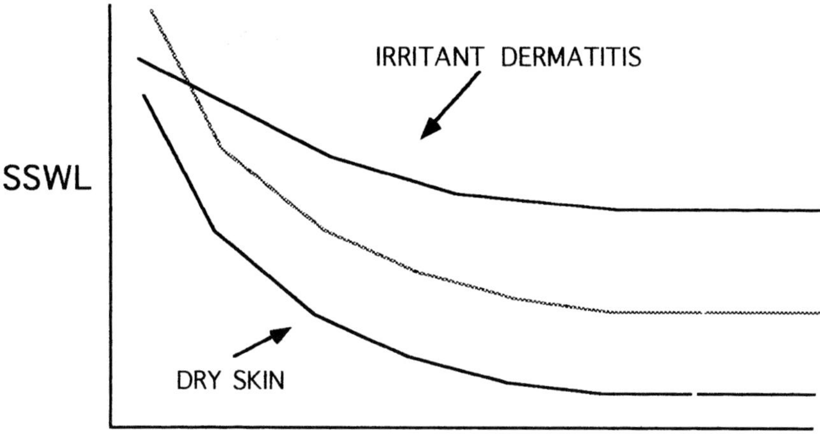

FIGURE 2   Differentiation using the POST between constitutional dry skin and subclinical cronic irritant dermatitis (see text).

bound water compartment is interested by evaporation (flat curve, second exponential).[10,11]

Pathological conditions of the skin may change the shape and the slope of the curve (Figure 2). A flat curve, with high SSWL levels during all the deconvolution time associated with high TEWL levels at the end, is detectable during irritant dermatitis and is due to the increase of TEWL secondary to barrier function impairment.[12] In this case, TEWL (barrier damage) exceeds SSWL values. On the contrary, dry skin shows a curve with lower SSWL levels throughout the deconvolution time.

The quality of the occlusion and the material used to induce the occlusion may influence the final results. Recently, some of us have compared, using the POST technique, the occlusive potential on healthy skin of plastic chamber, hydrocolloid, polyurethane, and polyethylene dressings:[13] the slopes are equal for all materials tested, but skin hydration and occlusive potential are higher for plastic chamber. This finding has a clinical relevance since occlusion has an important role in controlling epidermal turnover, wound healing, and microbial flora.[14] Therefore the POST can help in discriminating between the therapeutic efficiency of different dressing according to their occlusive potential.

The POST has also been used to investigate and quantify visible and nonvisible skin irritation induced by surfactants, to measure functional properties of the stratum corneum in dry pathological skin, and to assess the influence of some classes of lipids on the stratum corneum WHC.

Four surfactants (sodium lauryl sulfate [SLS], benzalkonium chloride, sorbitan monolaurate, and tegobetaine) belonging to different classes and widely utilized in manufacturing skin products have been compared using the POST technique to quantify the aggressive potential of these products.[15] The surfactants have been applied on skin surface in occlusion for 24 h at a

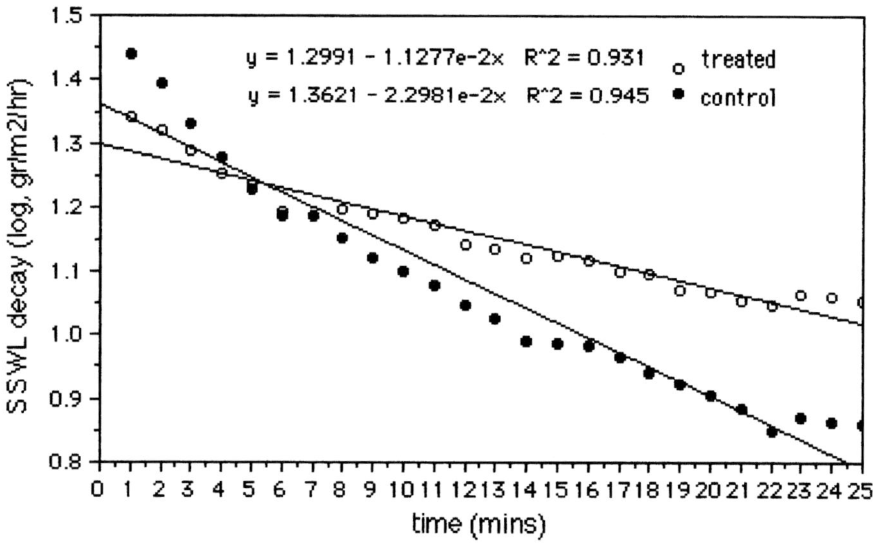

FIGURE 3   Effect of nonvisible irritation on skin surface water loss using the POST technique. Deconvolution curves show lower hydration (first minutes) and higher barrier damage (terminal part of the curve) in treated site (white dots).

concentration commonly used in the literature to induce skin irritation. After 24 h the plastic occlusion using plastic chamber has been applied on each irritated site. The POST decay curves showed a flat slope recorded for SLS compared to the other chemicals and control skin suggesting that SLS is the most aggressive surfactant and, therefore is a good model to induce and study irritant reactions.

Chronic irritant dermatitis has been investigated using the SLS-induced irritation model. Nonvisible dermatitis has been elicited by open SLS application to simulate routine use.[12] The application was repeated for 3 days and did not result in visible irritation, but significant changes between the treated and untreated sites were detected. The results (transformed in log) are shown in Figure 3. The slope of the treated site is less steep compared to the control, with lower Y intercept (drier skin) and higher ending part (increased barrier damage). Despite the lower water content at the first minute after the removal of the occlusion the SSWL decay is higher in the treated site because the surfactant-induced increase of TEWL biases, after 5 min, the decay curve: in this way TEWL becomes the prevalent constituent of skin evaporation. The study revealed that surfactant-induced changes are present just after a few nonocclusive applications even though the skin appears visually normal from a clinical and subjective viewpoint. The POST provides useful information not only on TEWL, but also on stratum corneum water content and water kinetics.

The usefulness of the POST in investigating nonvisible changes in skin function has been further assessed in the evaluation of stratum corneum WHC, hydration, and barrier function in apparently normal skin in psoriasis and

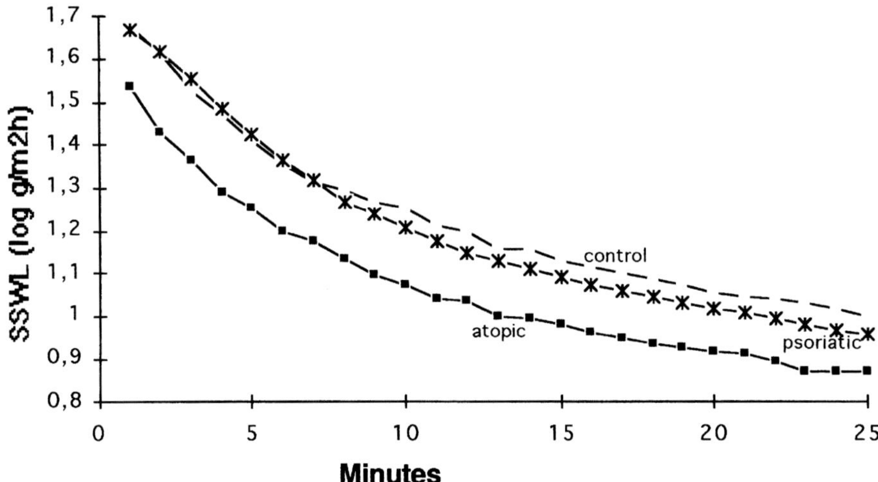

FIGURE 4   POST deconvolution curves in uninvolved skin of psoriatics and atopics. (From Faegermann et al.[9])

atopic dermatitis.[10] Subjects with no clinical signs of disease and/or skin dryness entered the study. Measurements (basal recordings and POST) were performed on two sites of the volar forearm. The decay curves in psoriatics and atopics were compared to controls (Figure 4): no significant differences were detected between uninvolved psoriatic skin and controls: however, while evaporation of free water is similar in the two groups, the bound water compartment shows reduced hydration in psoriatics due to the defective keratinization and stratum corneum formation present in apparently normal skin of psoriatic subjects. Therefore, the decay constants show a higher rate of decay in psoriasis, meaning a decreased WHC. The atopic curve shows a shape similar to control skin, but with lower SSWL values, corresponding to a drier skin condition. Indeed, basal TEWL and conductance values[10] indicate that the atopic group differs from controls and psoriatics being characterized by high TEWL and low water content values. The slope is not different from controls, thus we can assume that WHC in these two groups is similar. The data suggest that clinically normal skin may be functionally abnormal, resulting in a defective barrier that could lead to a higher risk for contact dermatitis.

Skin lipids have a critical role in maintaining the efficiency of the barrier function and the water-holding properties of the stratum corneum. Exposure to lipid solvents has been shown *in vitro* to reduce the water-holding properties of the stratum corneum;[16] *in vivo* studies are restricted by the irritant effects of solvents on skin and by environmental factors, which interfere with measurements. Thus, the limits of the POST to investigate stratum corneum WHC have been evaluated in a recent study:[17] the effects of two solvent mixtures have been investigated after short-term application on the skin in terms of skin hydration and stratum corneum WHC. In this study skin lipids have been

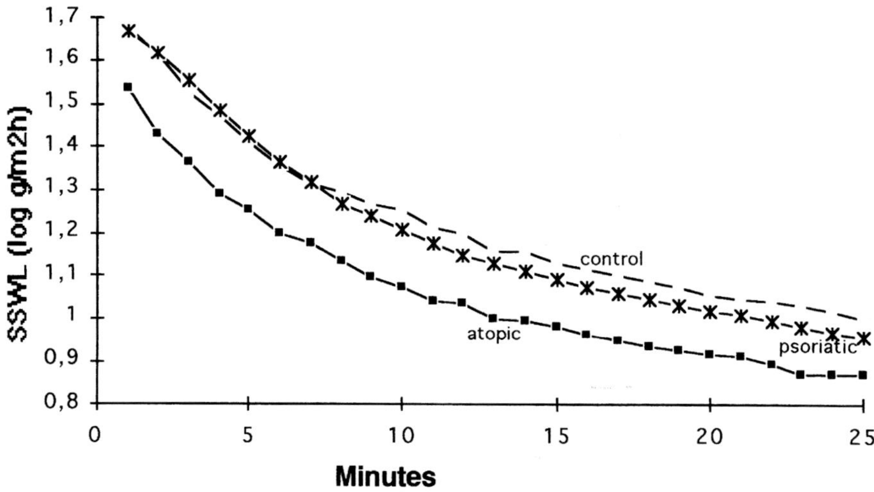

FIGURE 5 Skin surface water loss (SSWL) after removal of occlusion in delipidized and control sites. Stratum corneum water is significantly decreased in CM-treated compared to EA-treated and control sites from the fifth minute onward.

extracted by applying for 1 min ether/acetone (1:1)(E/A) or chloroform/methanol (2:1)(C/M). The first mixture extracts sebaceous gland lipids, cholesterol, and free fatty acids, while the second one extracts lipids more involved in the mechanism of water binding such as ceramides.

Figure 5 shows reduced evaporation of water in the site treated with C/M consequent to a reduced amount of water bound in the corneum. From the tenth minute onward, C/M results in significant lower WHC levels compared to control site ($p < 0.01$) and E/A ($p < 0.05$). In this case SSWL, which differs from TEWL, is lower after removal of lipids because a lower amount of bound water is available for evaporation. Indeed C/M extraction removes ceramides and other polar lipids such as sphingomyelin; furthermore, C/M also extracts free amino acids, which have been considered to have a role in maintaining stratum corneum WHC and skin suppleness.[18]

The POST thus seems a good method to measure WHC, hydration, and barrier function of the skin, and to estimate the free and bound water compartments. The main disadvantages are represented by the difficult mathematical approach and the analysis of the decay curves and decay constants, by the heating of the probe that can result in an increased level of recordings, and the time required to measure one single spot in each patient and in analyzing the data. We hope, however, that this method could be further improved, especially in computer software, to speed up the procedure and give more precise and reliable data for better *in vivo* evaluation of skin function.

References

1. **Elias, P.M.,** Epidermal lipids, barrier function and desquamation, *J. Invest. Dermatol.,* 80, 44, 1983.
2. **White, S.H., Mirejovsky, P., and King, G.I.,** Structure of lamellar lipid domains and corneocyte envelopes of murine stratum corneum, *Biochemistry,* 227, 3725, 1988.
3. **Lampe, M.A., Burlingame, A.L., Whitney, J., Williams, M.L., Brown, B.E., Roitman, E., and Elias, P.M.,** Human stratum corneum lipids: characterization and regional variations, *J. Lipid. Res.,* 24, 120, 1983.
4. **Elias, P.M., Cooper, E.R., Korc, A., and Brown, B.E.,** Percutaneous transport in relation to stratum corneum structure and lipid composition, *J. Invest. Dermatol.,* 76, 297, 1981.
5. **Nilsson, G.E.,** Measurement of water exchange through the skin, *Med. Biol. Eng. Comput.,* 15, 209, 1977.
6. **Agner, T. and Serup, J.,** Skin reactions to irritants as assessed by non-invasive bioengineering methods, *Contact Derm.,* 20, 352, 1989.
7. **Rietschel, R.L.,** A method to evaluate skin moisturizers in vivo, *J. Invest. Dermatol.,* 70, 152, 1978.
8. **Aly, R. and Maibach, H.,** Effects of prolonged skin occlusion, in *Neonatal Skin: Structure and Function,* Maibach, H.I. and Boisitis, E.K., Eds., Marcel Dekker, New York, 1983.
9. **Faegermann, J., Aly, R., Wilson, D.R., and Maibach, H.I.,** Skin occlusion: effect on pityrosporum orbiculare, Skin P_{CO_2}, pH, transepidermal water loss and water content, *Arch. Dermatol. Res.,* 275, 383, 1983.
10. **Berardesca, E., Fideli, D., Borroni, G., Rabbiosi, G., and Maibach, H.,** In vivo hydration and water-retention capacity of stratum corneum in clinically uninvolved skin in atopic and psoriatic patients, *Acta Dermatol. Venereol. (Stockholm),* 70, 400, 1990.
11. **Werner, Y., Lindberg, M., and Forslind, B.,** The water-binding capacity of stratum corneum in dry non-eczematous skin of atopic eczema, *Acta Dermatol. Venereol. (Stockholm),* 62, 334, 1982.
12. **Berardesca, E. and Maibach, H.I.,** Monitoring the water holding capacity in visually nonirritated skin by plastic occlusion stress test (POST), *Clin. Exp. Dermatol.,* 15, 107, 1990.
13. **Berardesca, E., Vignoli, G.P., Fideli, D., and Maibach, H.,** Effect of occlusive dressings on the stratum corneum water holding capacity, *Am. J. Med. Sci.,* 304, 25, 1992.
14. **Berardesca, E. and Maibach, H.,** Skin occlusion: treatment or drug-like device? *Skin Pharmacol.,* 1, 207, 1988.
15. **Berardesca, E., Fideli, D., Gabba, P., Cespa, M., Rabbiosi, G., and Maibach, H.I.,** Ranking of surfactant skin irritancy in vivo in man using the plastic occlusion stress test (POST), *Contact Derm.,* 23, 1, 1990.

16. **Yamamura, T. and Tezuka, T.,** The water holding capacity of the stratum corneum measured by 3H-NMR, *J. Invest. Dermatol.*, 93, 160, 1989.
17. **Berardesca, E., Herbst, R., and Maibach, H.,** The plastic occlusion stress test as a model to investigate the effects of skin delipidization on the stratum corneum water holding capacity in vivo, *Dermatology*, 187, 91, 1993.
18. **Imokawa, G., Kuno, H., and Kawai, M.,** Stratum corneum lipid serve as a bound water modulator, *J. Invest. Dermatol.*, 96, 845, 1991.

13 Freund's Adjuvant-Induced Skin Lesions in Rats: A Novel Subchronic Model for Assessing Antiinflammatory Agents

Gerard J. Gendimenico and James A. Mezick

TABLE OF CONTENTS

I. INTRODUCTION

A number of dermatological skin conditions exhibit inflammation and are chronic in nature. Among the most common of these disorders are psoriasis and atopic dermatitis. Although the etiology of these two disorders is still being elucidated, current evidence suggests that disturbances in the immune system are key components of their pathophysiology.[1-3]

For many human diseases, animal models are useful not only for understanding underlying pathology, but also serve as a means to identify potential therapeutic agents. In dermatology, there has been a continual need for relevant animal models of skin disease. A number of animal models of skin inflammation have been described,[4] however, most of these are of acute duration and only partly mimic aspects of human diseases. Another drawback of these models is that pharmacological agents sometimes exert different effects in animals compared to human skin diseases.

In this chapter, we describe a novel subchronic model of skin inflammation in rats that is induced by injecting complete Freund's adjuvant (oil suspension of dead *Mycobacterium*) into the footpad. As a result of this treatment, the animals develop polyarthritis and inflamed skin lesions on the ears and tail. The polyarthritis is widely recognized and has been useful as a model for human rheumatoid arthritis.[5] Although the histopathology of the skin lesions in adjuvant-treated rats was first described by Pearson et al.[6] over 30 years ago, the nature of the skin lesions has not been fully investigated. We have therefore evaluated this model for its potential utility as a relevant test for topical antiinflammatory agents.

II. METHODOLOGY

Skin lesions were induced with Freund's adjuvant in male Lewis strain rats (150 to 175 g, Charles River Laboratories). A 7.5 mg/ml suspension of *Mycobacterium butyricum* (heat killed and dried, Difco, Detroit, MI) was prepared using light mineral oil. A 0.1 ml volume of this suspension was injected into the subplantar tissue of the right hind paw. Pharmacological agents were prepared in an ethanol–propylene glycol (70:30, v/v) vehicle. A 20 μl volume of vehicle or drug solution was applied to each ear with both ears of each rat receiving the same treatment. The effects of drug treatment were quantified by counting the number of lesions on each ear. A count of the total number of lesions on days 14 or 15 prior to the first treatment was used to establish a baseline. The percent change in the number of lesions during treatment was calculated relative to the day 14 or 15 lesion number. Lesions were microscopically evaluated using hematoxylin and eosin-stained vertical sections prepared from formalin-fixed ears.

III. CLINICAL OBSERVATIONS AND HISTOPATHOLOGY

No skin lesions are observed until 13 to 15 days after adjuvant administration. Although 100% of the animals show arthritic abnormalities, approximately 80% of animals exhibit skin lesions on the ear and tail. Ear lesions are favored over tail lesions for study because they can be easily quantified by visual counting. These lesions, occurring on both left and right ears, are 2 to 3 mm in size with a focal, erythematous, elevated appearance. On average, one to three lesions per ear are present on days 14 or 15 after adjuvant treatment.

Histological evaluation shows lesions that exhibit pronounced epidermal hyperplasia and marked infiltration of inflammatory cells.[7] In the epidermis, neutrophils are present between keratinocytes and on the surface of the stratum corneum, whereas in the dermis, cells are primarily lymphocytic. Dilated blood vessels are often present in lesional skin, which accounts for the intense erythema exhibited by lesions.

IV. EFFECTS OF ANTIINFLAMMATORY AGENTS

Glucocorticoids, cyclosporine, and indomethacin were examined for their topical effects on developed skin lesions. Topical glucocorticoids are the most widely used antiinflammatory drugs in dermatology and are effective in a number of skin diseases.[8] Cyclosporine, an immunosuppressive agent, has shown marked efficacy in psoriasis and atopic dermatitis after oral administration, but is inactive in these two diseases when used topically.[2,9-11] Indomethacin is generally ineffective as an antiinflammatory drug in chronic inflammatory skin disease and, moreover, has been shown to exacerbate psoriasis.[12]

Compounds were applied once daily for up to 7 days and the lesions were counted at various intervals. Table 1 shows that two glucocorticoids, dexamethasone and fluocinolone acetonide, caused almost a complete reduction of lesions after 7 days of dosing. Significant effects were observed as early as after two daily treatments. In the vehicle control group the number of skin lesions did not change greatly during the 7-day dosing period. Cyclosporine at 3%, administered in the same fashion as the glucocorticoids, significantly reduced the number of skin lesions at 20 and 22 days after adjuvant administration.[7] The effects of the glucocorticoids and cyclosporine are clearly dose-related (Figure 1). The glucocorticoids are more potent than cyclosporine at reducing skin lesions.

Topical indomethacin has effects opposite to those of glucocorticoids and cyclosporine. Table 2 shows that indomethacin caused a time and dose-dependent exacerbation of skin lesion numbers.

TABLE 1
Time-Course of Skin Lesion Reduction by Topical Glucocorticoids

Treatment	Dose (%)	Reduction of skin lesions vs. day 14 (%)		
		Day 17	Day 19	Day 22
Vehicle	—	−18	−6	6
Fluocinolone acetonide	0.005	39[a]	64[a]	86[a]
Dexamethasone	0.1	35[a]	70[a]	95[a]

Note: Ears were treated for seven consecutive days, on days 15 through 21, after adjuvant administration and lesions quantified at various intervals. $N = 8$ rats per group.

[a] Significantly different ($p < 0.05$) from corresponding vehicle treatment (analysis of covariance).

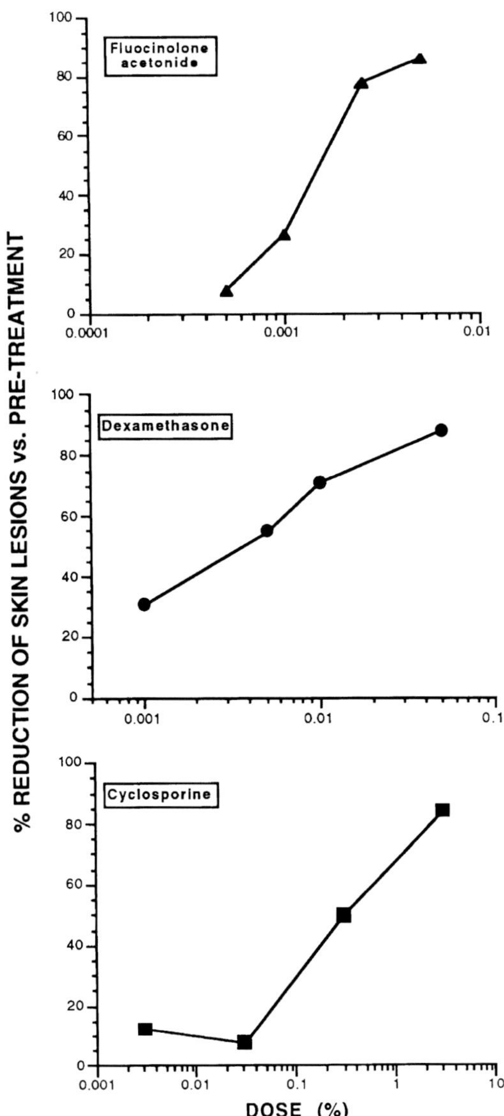

FIGURE 1 Dose-related clearing of skin lesions by topical glucocorticoids (fluocinolone acetonide, dexamethasone) and cyclosporine. Ears were treated for seven consecutive days, on days 15 through 21, after adjuvant administration and lesions quantified on day 22. Baseline pretreatment skin lesion counts were taken on day 14 for fluocinolone acetonide and day 15 for dexamethasone and cyclosporine.

TABLE 2
Enhancement of Skin Lesions by Topical Indomethacin

		Change in skin lesions vs. day 15[a] (%)		
Treatment	Dose (%)	Day 18	Day 20	Day 22
Vehicle	—	13	9	−17
Indomethacin	0.1	68	44	32
	1.0	65	96[b]	127[b]

Note: Ears were treated for seven consecutive days, on days 15 through 21, after adjuvant administration and lesions quantified at various intervals. $N = 8$ rats per group.

[a] Positive values indicate an increase in the number of skin lesions.

[b] Significantly different ($p < 0.05$) from corresponding vehicle treatment (analysis of covariance).

V. DISCUSSION AND CONCLUSIONS

Our findings with adjuvant-skin lesions in rats show it to be a useful model for assessing the efficacy of dermatotherapeutic antiinflammatory compounds. In comparison to existing acute cutaneous inflammation models, the adjuvant-induced skin lesion model offers some advantages. The drug treatments are given after the development of lesions and simulate clinical conditions where drugs are used in a therapeutic regimen rather than prophylactically. The number of lesions remain sufficiently elevated for at least 7 days so that treatment can be administered multiple times. Another advantage of the adjuvant model is that skin lesions appear spontaneously after a single injection of adjuvant into the footpad. This is unlike previously described models that require an inflammatory stimulus to be applied or injected directly to skin sites destined for treatment.[4] To induce subchronic inflammation with exogenous stimulants, it is necessary to apply stimuli multiple times.[13]

The response of the lesions to topical drug treatment was similar to how the drugs behave in psoriasis: clearing by glucocorticoids and cyclosporine, but exacerbation by indomethacin. We found cyclosporine active topically at clearing lesions even though it is ineffective by the topical route in psoriasis and atopic dermatitis. This difference is probably due to the greater degree of penetration of cyclosporine through rat skin compared to human skin.[11] Cyclosporine has been reported to have topical antiinflammatory activity in allergic contact dermatitis in guinea pigs.[14] The exacerbation of skin lesions by indomethacin is a novel response in this model because in other cutaneous inflammation models, indomethacin either reduces inflammation or is without effect.[13,15-18] The mechanism of exacerbation, although uncertain, may be due to loss of immunosuppressive prostaglandins[19] or enhanced production of proinflammatory leukotrienes.

In summary, we show here that Freund's adjuvant-induced skin lesions represent a novel and potentially useful model for assessing agents that modulate skin inflammation. It is known that an arthritic pathology can be induced in normal rats by passive transfer of lymphocytes from adjuvant-treated rats, thus implicating a cell-mediated mechanism for adjuvant-induced disease.[20,21] Thus, this model appears especially relevant for human skin diseases that are immune-mediated and could be particularly useful for understanding the pathophysiology of human inflammatory skin conditions, such as psoriasis and atopic dermatitis.

References

1. **Griffiths, C.E.M. and Voorhees, J.J.,** Immunological mechanisms involved in psoriasis, *Springer Semin. Immunopathol.*, 13, 441, 1992.
2. **Cooper, K.D.,** Atopic dermatitis: recent trends in pathogenesis and therapy, *J. Invest. Dermatol.*, 102, 128, 1994.
3. **Wong, R.L., Winslow, C.R., and Cooper, K.D.,** The mechanisms of action of cyclosporin A in the treatment of psoriasis, *Immunology Today*, 14, 69, 1993.
4. **Young, J.M. and De Young, L.M.,** Cutaneous models of inflammation for the evaluation of topical and systemic pharmacological agents, in *Pharmacological Methods in the Control of Inflammation*, Chang, J.Y. and Lewis, A.J., Eds., Alan R. Liss, New York, 1989, 215.
5. **Weichman, B.M.,** Rat adjuvant arthritis: a model of chronic inflammation, in *Pharmacological Methods in the Control of Inflammation*, Chang, J.Y. and Lewis, A.J., Eds., Alan R. Liss, New York, 1989, 363.
6. **Pearson, C.M., Waksman, B.H., and Sharp, J.T.,** Studies of arthritis and other lesions induced in rats by injection of mycobacterial adjuvant. V. Changes affecting the skin and mucous membranes: comparison of the experimental process with human disease, *J. Exp. Med.*, 113, 485, 1961.
7. **Gendimenico, G.J. and Mezick, J.A.,** Effects of topical antiinflammatory agents on Freund's adjuvant-induced skin lesions in rats, *Agents Actions*, 44, 16, 1995.
8. **Chren, M.M. and Bickers, D.R.,** Dermatological pharmacology, in *The Pharmacological Basis of Therapeutics*, 8th Ed., Gilman, A.G., Rall, T.W., Nies, A.S., and Taylor, P., Eds., Pergamon Press, New York, 1990, 1572.
9. **Ellis, C.N., Gorsulowsky, D.C., Hamilton, T.A., Billings, J.K., Brown, M.D., Headington, J.T., Cooper, K.D., Baadsgaard, O., Duell, E.A., Annesley, T.M., Turcotte, J.G., and Voorhees, J.J.,** Cyclosporine improves psoriasis in a double-blind study, *JAMA*, 256, 3110, 1986.
10. **Sowden, J.M., Berth-Jones, J., Ross, J.S., Motley, R.J., Marks, R., Finlay, A.Y., Salek, M.S., Graham-Brown, R.A.C., Allen, B.R., and Camp, R.D.R.,** Double-blind, controlled, crossover study of cyclosporin in adults with severe refractory atopic dermatitis, *Lancet*, 338, 137, 1991.

11. **Hermann, R.C., Taylor, R.S., Ellis, C.N., Williams, N.A., Weiner, N.D., Flynn, G.L., Annesley, T.M., and Voorhees, J.J.,** Topical ciclosporin for psoriasis: In vitro skin penetration and clinical study, *Skin. Pharmacol.*, 1, 246, 1988.

12. **Ellis, C.N., Fallon, J.D., Kang, S., Vanderveen, E.E., and Voorhees, J.J.,** Topical application of nonsteroidal anti-inflammatory drugs prevents vehicle-induced improvement of psoriasis, *J. Am. Acad. Dermatol.*, 14, 39, 1986.

13. **Stanley, P.L., Steiner, S., Havens, M., and Tramposch, K.M.,** Mouse skin inflammation induced by multiple topical applications of 12-O-tetradecanoylphorbol-13-acetate, *Skin Pharmacol.*, 4, 262, 1991.

14. **Nakagawa, S., Oka, D., Jinno, Y., Takei, Y., Bang, D., and Ueki, H.,** Topical application of cyclosporine on guinea pig allergic contact dermatitis, *Arch. Dermatol.*, 124, 907, 1988.

15. **Carlson, R.P., O'Neill-Davis, L., Chang, J., and Lewis, A.J.,** Modulation of mouse ear edema by cyclooxygenase and lipoxygenase inhibitors and other pharmacologic agents, *Agents Actions*, 17, 197, 1985.

16. **Griffiths, R.J., Wood, B.E., Li, S., and Blackham, A.,** Pharmacological modification of 12-O-tetradecanoylphorbol-13-acetate induced inflammation and epidermal cell proliferation in mouse skin, *Agents Actions*, 25, 344, 1988.

17. **De Young, L.M., Kheifets, J.B., Ballaron, S.J., and Young, J.M.,** Edema and cell infiltration in the phorbol ester-treated mouse ear are temporally separate and can be differentially modulated by pharmacologic agents, *Agents Actions*, 26, 335, 1989.

18. **Meurer, R., Opas, E.E., and Humes, J.L.,** Effects of cyclooxygenase and lipoxygenase inhibitors on inflammation associated with oxazolone-induced delayed hypersensitivity, *Biochem. Pharmacol.*, 37, 3511, 1988.

19. **Goodwin, J.S. and Webb, D.R.,** Regulation of the immune response by prostaglandins, *Clin. Immunol. Immunopathol.*, 15, 106, 1980.

20. **Taurog, J.D., Argentieri, D.C., and McReynolds, R.A.,** Adjuvant arthritis, *Methods Enzymol.*, 162, 339, 1988.

21. **Cohen, I.R.,** Regulation of autoimmune disease physiological and therapeutic, *Immunol. Rev.*, 94, 5, 1986.

14 A Model to Study the Drying Potential of Detergent Formulations on the Skin

Pieter G.M. van der Valk, Els B. Stam-Westerveld, and Marc Paye

TABLE OF CONTENTS

I. INTRODUCTION

Many people complain about dry skin, which refers to the dry appearance of its surface. Although dry skin is difficult to define, the incidence of dry skin correlates with an increased skin reactivity to irritants.[1]

Dry skin may be observed in atopic dermatitis, in the ichthyoses, and in the elderly and may be a mild form of asteatotic eczema.[2] Skin exposure to mild irritants has also been incriminated and dry skin, as caused by irritants, may be defined as low-grade irritant contact dermatitis.[3] The skin comes into contact with a lot of chemicals, most frequently with body cleansers and skin care products. Detergent-based products, if not designed to respect the stratum corneum, may be particularly irritating. Detergents may also defat the skin, which may also contribute to a dry appearance of the skin.[4,5] Irritation includes observable signs, such as erythema and skin flaking or wrinkling, as well as self-perceived effects, such as stinging and the feeling of dryness.[6]

Skin care attitudes have changed in modern society, and, as such, body cleansers are used more intensively than a few decades ago. Millions of consumers use body cleansers daily, of which a substantial minority complains of skin irritation and dry skin.[7] This puts a larger responsibility on the manufacturer to develop mild products, i.e., products that do not cause irritant contact dermatitis or dry skin even if they are used once or twice daily.

A lot of research has been performed to study the drying effect of detergents.[8,9] Most models do not realistically mimic the daily use situation, and use tests are very difficult to perform because of confounding factors such as differences in the skin's exposure to irritants and differences in skin vulnerability among panelists. Therefore, *in vivo* models simulating daily use of detergents, in which confounding factors can be avoided or corrected for, may be useful. Many of the routine *in vivo* screening methods that have been developed to assess irritation and dryness utilize prolonged exposures (24 h and above) under occlusive patches.[10,11] These methods mainly induce erythema with a little dryness, which develops consecutively to erythema after the exposure to the irritant is complete. These prolonged exposures do not reflect the daily use situation.

In normal life, a person's skin is exposed to the irritant for a brief period several times a day under nonoccluded conditions.

In this chapter, we describe a model that evaluates the ability of surfactant-based products to induce dryness with repeated short-term exposures using open patches and that discriminates between products in their drying potential. The concentration of the test solutions and the exposure time and frequency were selected to cause only slight erythema reactions in most subjects simulating the development of dry skin from daily use. The dorsal side of the lower leg is used because the legs are a predilection side for experiencing dry skin.[12] Highly sensitive people were selected (volunteers with a history of atopic dermatitis).

II. MATERIALS AND METHODS

All substances were tested simultaneously (within-subject comparison) and double-blind. The substances were applied to the dorsal side of the lower legs along the longitudinal axis in a latin square. An open test procedure was used: at both sides open polypropylene cylinders with a inner diameter of 22 mm attached in a thin copper plate are fixed to the skin with nonadhesive tape.

The skin was exposed for 30 min twice daily for five consecutive days with at least 2 h in between. After the exposure the plate was removed and the test sites were thoroughly rinsed with running tap water. If intense irritation (score of 3 or above on the visual erythema scale) should occur the applications were stopped for all products and data from those panelists not considered anymore for later assessments.

No statistical analysis was performed due to the low panel size.

A.   TEST SUBSTANCES

Marketed body cleansing materials were used and supplied under code one to three. Products were coded as follows: product 1 = facial cleanser; product 2 = soap; product 3 = synthetic detergent bar. Sodium dodecyl sulfate (SDS) was used as a positive control and distilled water as a negative control. The cleansing materials were tested as 10% w/v and SDS in 1% w/v in distilled water.

B.   TEST SUBJECTS

Ten healthy caucasian volunteers, of both sexes, ranging in age from 20 to 31 years with a history of atopic dermatitis were included after giving informed consent. Panelists were free of cutaneous symptoms on their lower legs that could interfere with scoring. During the study, panelists were not allowed to use moisturizers or soaps on their lower legs and were asked to avoid activities that could interfere with the test.

C.   VISUAL ASSESSMENTS

The scoring was done prior to the first product application, 30 min after the second exposures on days 3 and 5, as well as 3 days after last exposure (day 8).

Clinical assessments were made by a trained evaluator based on standardized scales for erythema and dryness (scaling and wrinkling) as follows:

Erythema: 0 = no redness
           1 − slight redness, spotty or diffuse
           2 = moderate confluent redness
           3 = intense confluent redness
           4 = fiery red

Scaling: 0 = no dryness
           1 = fine flaking
           2 = moderate flaking/scaling
           3 = severe flaking

Wrinkling: 0 = completely smooth relief
           1 = slightly irregular relief
           2 = slight wrinkling
           3 = moderate, palpable wrinkling
           4 = strong palpable wrinkling

D.   INSTRUMENTAL

Transepidermal water loss was measured using an evaporimeter (ServoMed, Stockholm, Sweden) according to the guidelines for transepidermal water loss (TEWL) measurement by the Standardization Group of the European Society of Contact Dermatitis.[13]

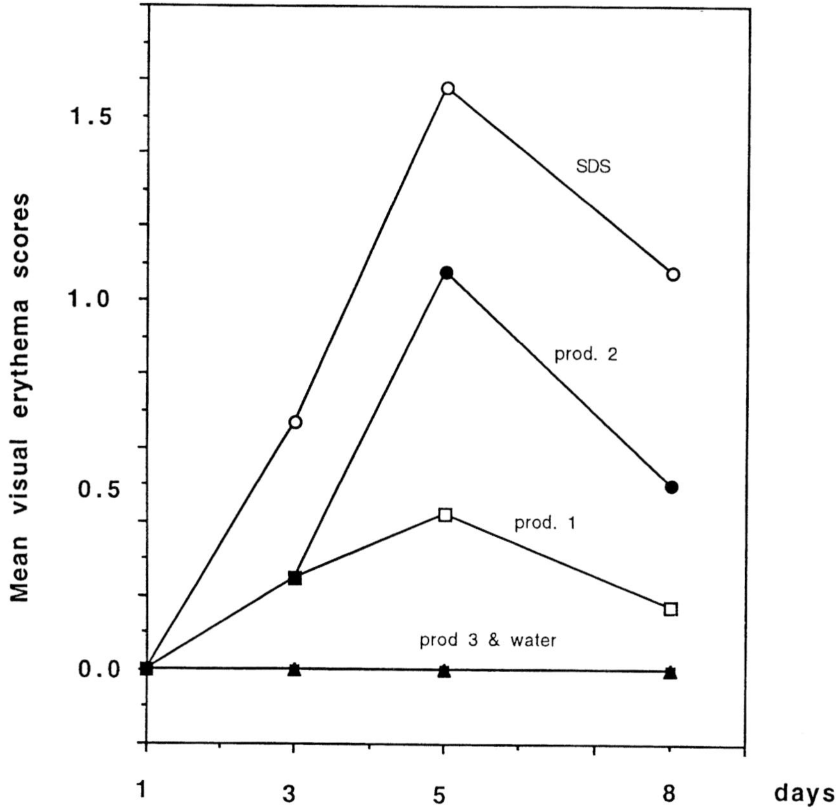

FIGURE 1   Erythema after multiple 30-min exposures to cleansing formulations and SDS.

Skin color was measured instrumentally using a Chromameter (Minolta, Osaka, Japan) as described by Babulak et al.[14] Skin conductance was measured using a Corneometer (Courage & Khazaka, Cologne, Germany) and related to skin dryness.[15] D-Squame® tapes (CuDerm Corp., Dallas, TX) were taken from the test sites on day 8. Evaluation of the D-Squame® tapes was done according to a four-level reference scoring system for dryness: 1+ = normal, 2+ = low dryness, 3+ = moderate dryness, 4+ = severe dryness.[16]

III. RESULTS

Slight to moderate erythema was observed for products 1 and 2, respectively. Product 3 and water did not induce erythema. SDS is a well-known skin irritating substance, which proved to be irritating in this model as well (Figure 1).

Chromameter a^* measurements showed color changes in accordance with visual scoring for products 1 and 2 as well as for SDS. In addition, slight color changes were measured for product 3 (Figure 2).

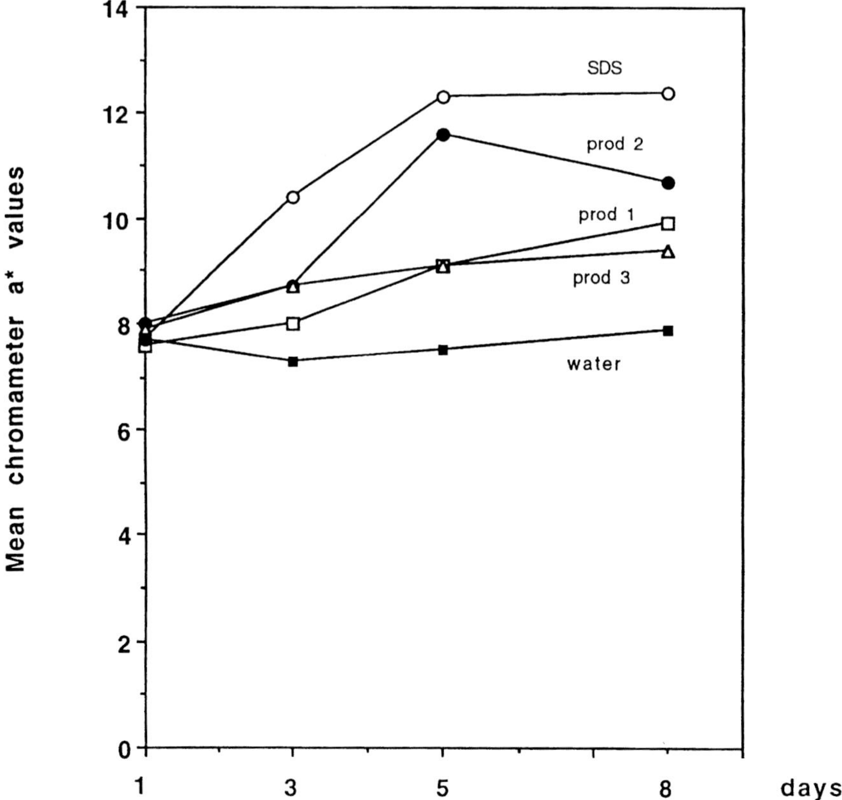

FIGURE 2   Skin color after multiple 30-min exposures to cleansing formulations and SDS.

Dryness, defined as the sum of scaling and wrinkling, mainly developed as wrinkling and was pronounced for SDS, less pronounced for products 1 and 2, and low for the other products at day 5. For the two solutions having induced the most erythema (SDS and product 2), skin wrinkling continued to evolve after the application period in contrast with the products having induced no or minimal erythema (products 3 and 1). For these products wrinkling regressed from day 5 to 8 (Figure 3).

At day 8, D-Squame® tapes were taken and showed most desquamation for SDS and product 2. The water-treated sites displayed the least desquamation and product 1 and 3 were intermediate (Figure 4).

Skin dryness due to product 2 and SDS could be substantiated by skin conductance measurements at day 8, 3 days after the end of the applications (Figure 5). During the application period, conductance measurements were probably overestimated[15] when high TEWL values were observed, as it was the case for SDS at days 3 and 5 and for product 2 at day 5 (Figure 6). Table 1 summarizes all results collected at day 8 with the standard error of the means.

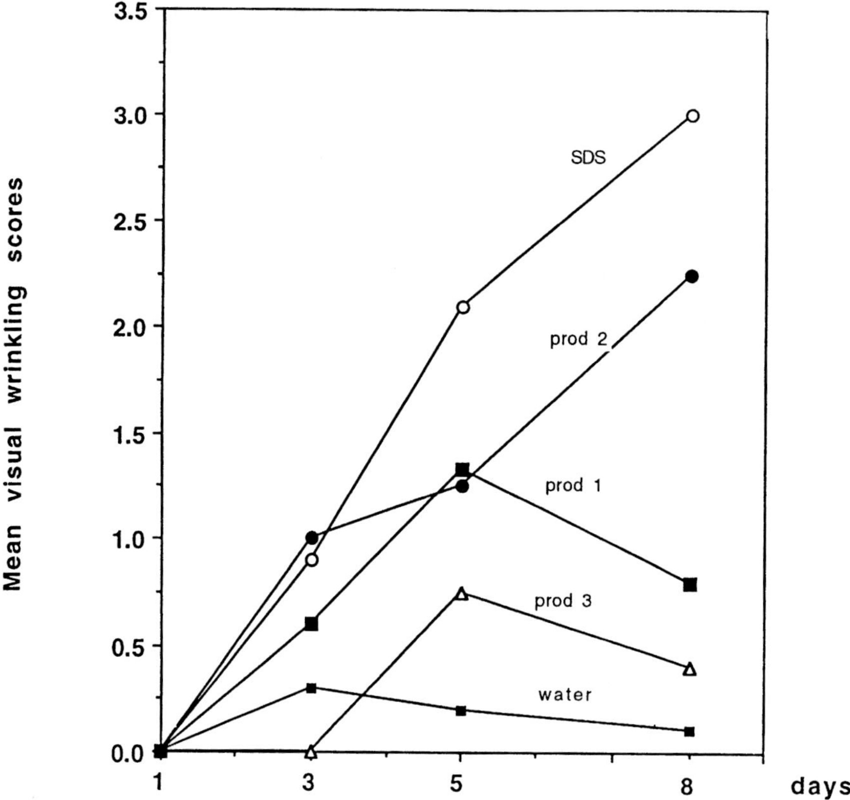

FIGURE 3 Wrinkling after multiple 30-min exposures to cleansing formulations and SDS.

The exposures were stopped at day 3 in two subjects because of excessive skin irritation by SDS. Data from these two subjects were not subsequently used at day 5 for any of the sites ($n = 8$ at day 5).

IV. DISCUSSION

This chapter presents a model to study *in vivo* skin irritation by surfactant-based products with several advantages over the classical occlusive patch test methodology.

Open patch tests are more realistic to daily product use than occlusive application. Indeed, under occlusion skin barrier properties become weaker, even in absence of irritating product; therefore, a product could interact with deeper skin layers, and hence display different types of reactions, than under in-use conditions. Also, under occlusion, the high humidity conditions under patch may delay the product-induced dryness.

Repeated short applications simulate normal product use better than single 24- or 48-h applications.

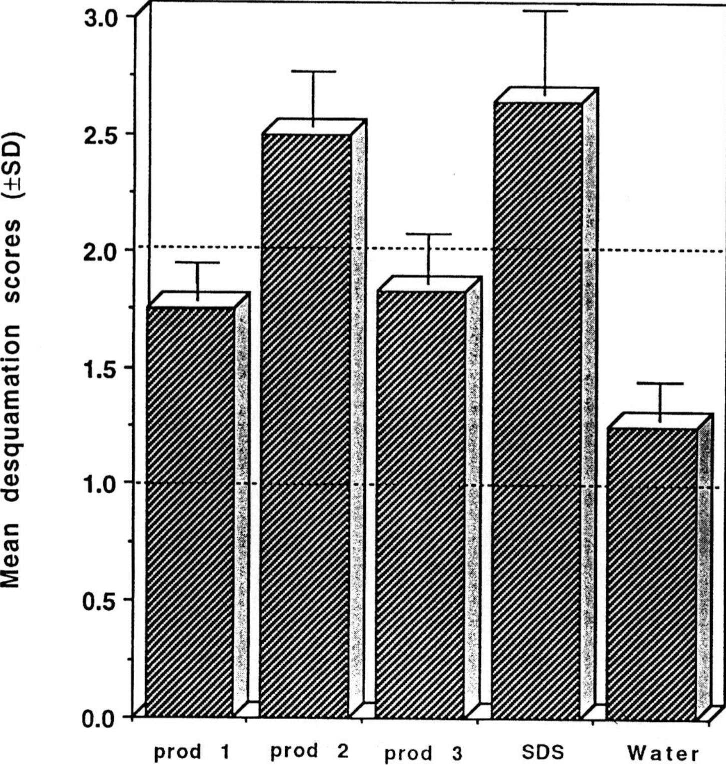

FIGURE 4 D-Squame® tapes taken at day 8 after multiple 30-min exposures to cleansing formulations and SDS.

In this model skin irritation develops progressively, mainly as skin dryness, as observed visually, by conductance measurements and by D-squames analysis, with limited erythematous reactions as compared to occlusive patch tests.

The data presented in this chapter suggest skin dryness from detergents develops according to two different kinetic pathways:

1. Skin dryness consecutive to an inflammatory reaction and continuing to develop after the application period as observed with SDS and product 2. This mode of development is similar to what is observed in occlusive patch tests where dryness usually appears consecutively to erythema. This dryness is related to the severity of the erythematous reaction. This is not what happens in daily life use when dryness appears before any clinical sign of erythema; and

2. Dryness caused by direct interaction of the product with the horny layer, with or without minimal inflammation as observed with products 1 and 3. This second type of dryness regresses immediately after stopping the application. In daily life, skin cleansing materials and skin care products, when inducing skin dryness, mostly do it in absence of inflammation.

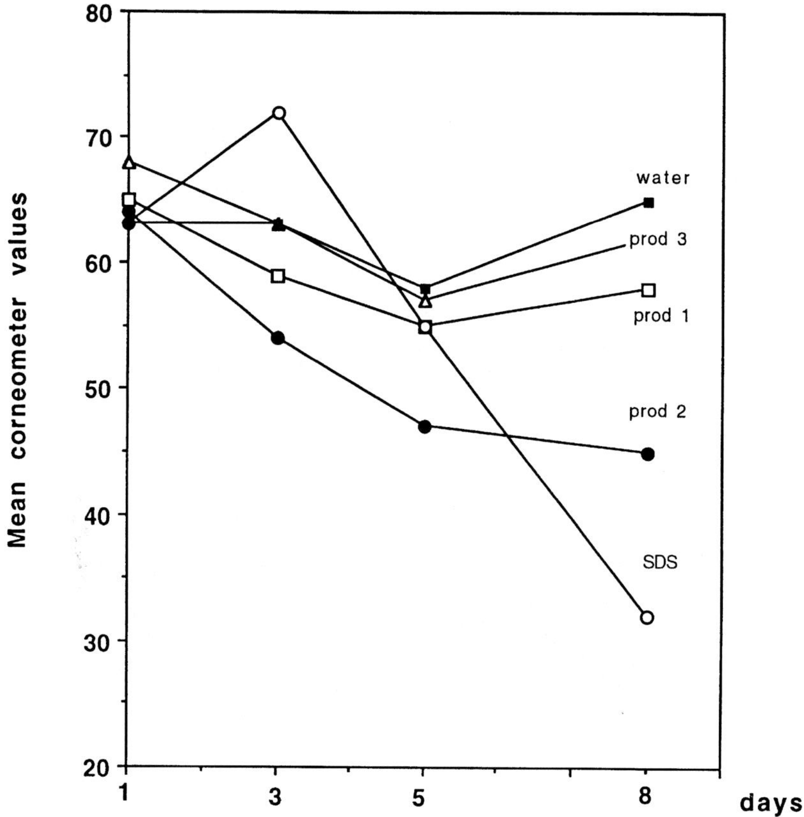

FIGURE 5 Conductance after multiple 30-min exposures to cleansing formulations and SDS.

The *in vivo* skin irritation model presented in this chapter provides a useful tool to study skin dryness induced by these later products in a mode similar to what is observed in daily life.

In the case of more irritating products such as SDS or product 2, the methodology should be adapted by reducing solutions concentration to investigate their direct drying effect on the horny layer in absence of erythema.

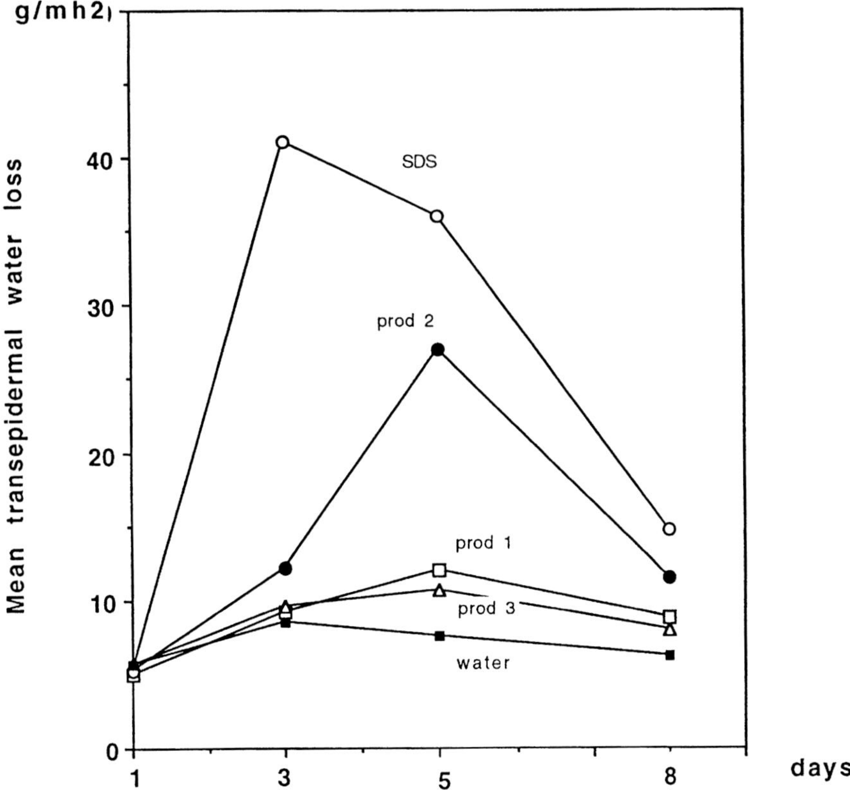

FIGURE 6   Transepidermal water loss (TEWL) after multiple 30-min exposures to cleansing formulations and SDS.

TABLE 1
Multiparameter Assessment of Skin Irritancy at Day 8 after Multiple 30 min Exposures to Cleansing Formulations and SDS (Mean ± SEM, $n = 8$)

	D-Squame tapes	Erythema	Chromameter a^*	Wrinkling	Conductance	TEWL
Product 1	1.75 ± 0.16	0.17 ± 0.14	9.87 ± 0.14	0.83 ± 0.29	57.67 ± 4.30	8.79 ± 1.26
Product 2	2.50 ± 0.24	0.50 ± 1.24	10.67 ± 0.53	2.25 ± 0.34	44.58 ± 4.73	11.53 ± 1.22
Product 3	1.83 ± 0.21	0	9.37 ± 0.40	0.42 ± 0.18	61.67 ± 3.60	7.87 ± 1.22
SDS	2.64 ± 0.36	1.08 ± 0.35	12.36 ± 0.48	3.00 ± 0.30	32.42 ± 5.56	14.71 ± 1.90
Water	1.25 ± 0.16	0	7.92 ± 0.28	0.08 ± 0.10	65.00 ± 4.18	6.23 ± 0.98
Baseline	Not determined	0	7.76 ± 0.16	0	64.88 ± 2.39	5.38 ± 0.33

References

1. **Tupker, R.A., Pinnagoda, J., Coenraads, P.J., and Nater, J.P.,** Susceptibility to irritants: role of barrier function, skin dryness and history of atopic dermatitis, *Br. J. Dermatol.*, 123, 199, 1990.

2. **Tupker, R.A.,** The influence of detergents on the human skin, Ph.D. Thesis, State University Groningen, The Netherlands, 1990, 31.

3. **Malten, K.E.,** Thoughts on irritant contact dermatitis, *Contact Derm.*, 7, 238, 1982.

4. **Idson, B.,** Dry skin: moisturizing and emolliency, *Cosmet. Toilet.*, 107, 69, 1992.

5. **Rieger, M.,** Skin lipids and their importance to cosmetic science, *Cosmet. Toilet.*, 102, 36, 1987.

6. **Frosch, P.J.,** Cutaneous irritation, in *Textbook of Contact Dermatitis*, Rycroft, R.J.G., Menné, T., Frosch, P.J., and Benezra, C., Eds., Springer-Verlag, Berlin, 1992, 28.

7. **Groot de, A.C.,** Side effects of cosmetics, Ph.D. Thesis, State University Groningen, The Netherlands.

8. **Kawai, M. and Okamoto, K.,** The influence of surfactants on the skin, *Jpn. J. Dermatol.*, 92, 465, 1982.

9. **Marks, R.,** Clinical methods for measuring scaling and desquamation, *Bioeng. Skin*, 3, 319, 1987.

10. **Philips, L., Steinberg, M., Maibach, H.I., and Akers, W.A.,** A comparison of rabbit and human skin responses to certain irritants, *Toxicol. Appl. Pharmacol.*, 21, 369, 1972.

11. **Frosch, P.J. and Kligman, A.M.,** The soap chamber test, *J. Am. Acad. Dermatol.*, 1, 35, 1979.

12. **Vilaplana, J., Coll, J., Trullás, Azon, A., and Pelejero, C.,** Clinical and noninvasive evaluation of 12% ammonium lactate emulsion for the treatment of dry skin in atopic and non-atopic subjects, *Acta Dermatol. Venereol. (Stockholm)*, 72, 28, 1992.

13. **Pinnagoda, J., Tupker, R.A., Agner, T., and Serup, J.,** Guidelines for transepidermal water loss (TEWL) measurement, *Contact Derm.*, 22, 164, 1990.

14. **Babulak, S.W., Rhein, L.D., Scala, D.D., Simion, F.A., and Grove, G.L.,** Quantification of erythema in a soap chamber test using the Minolta Chroma (Reflectance) Meter: comparison of instrumental results with visual assessment, *J. Soc. Cosmet. Chem.*, 37, 475, 1986.

15. **Paye, M., Van De Gaer, D., and Morrison, B.M., Jr.,** Skin dryness assessment in a soap chamber test: optimization of visual scoring and corneometer measurement, *Allergologie*, 4, 161, 1993.

16. **Kligman, A.M., Schatz, H., Manning, S., and Stoudemayer, T.,** Quantitative assessment of scaling in winter xerosis using image analysis of adhesive-coated disks (D-Squames®), in *Noninvasive Methods for the Quantification of Skin Functions*, Frosch, P.J. and Kligman, A.M., Eds., Springer-Verlag, Berlin, 1993, 309.

15 Some Aspects of *In Vitro* Percutaneous Penetration Studies Using Radioisotopes

Malgorzata Sznitowska

TABLE OF CONTENTS

At first glance, the technique of percutaneous absorption experiments conducted *in vitro* is very simple. An investigator cuts out a piece of skin, installs it in a diffusion chamber, then the compound under investigation is placed on the skin while a receiver fluid is pumped beneath the skin. In the fixed time periods samples of the receiver fluid are collected and concentration of the penetrant is analyzed.

Problems, however, arise, e.g., leakage of the fluid from the chamber, blistering or more serious damage of the skin, air bubbles, not to mention accidents such as rising temperature and disorders in the work of pump or stirrers. These are all technical problems, while the main scientific problem concerns the next phase of the experiment, i.e., analysis of the penetrant in the collected samples of the receiver fluid.

Analysis of many compounds in the presence of skin components extracted during the diffusion experiment is often very difficult. If a specific analytical reaction or technique may be employed, it usually means that analysis of the penetrant does not require an extraction step. Unfortunately, the necessity of this step is rather the rule. When the compound is lipophilic an extraction with an organic solvent is recommended, but for hydrophilic solutes a classic extraction is not applicable, because most of the compounds to be removed are also hydrophilic in nature. Chromatographic techniques, such as HPLC or GC, are of an invaluable help in such cases since they offer the possibility of separation and quantitative analysis, simultaneously.

The other most popular approach in percutaneous absorption studies is to use radioisotopes as tracers of the penetrant. It is true that the method is very sensitive and can deal with dilutions outside the range of most other methods. Besides that, this technique is believed to be easy, fast, and reliable. Some investigators find the procedure of looking for a radioisotope and purchasing it the most difficult step in the experiments. The intention of this chapter is to point out some problems, mainly dangers of misinterpretation of the results, while working with radiolabeled compounds.

For percutaneous absorption studies of organic compounds ^{3}H, ^{14}C, ^{35}S, ^{32}P (β-emitters) and ^{125}I isotopes (γ-emitter) are employed. Among them ^{3}H and ^{14}C are the most useful, and their applications will be discussed here.

Many of the radiochemicals are commercially available; others may be synthesized to order. Determination of radioactivity in the receiver fluid or in the skin is done by liquid scintillation counting. The liquid scintillation process is a simple and fast technique of radioactivity measurement. One of the most important attributes of this method is the ability to determine two isotopes simultaneously in the same sample. For this technique to be successful the β spectra of the two isotopes must be sufficiently different. This is the case for the isotope pairs most commonly used: ^{3}H and ^{14}C, ^{3}H and ^{35}S, and ^{3}H and ^{32}P. This makes it possible to determine in the same experiment skin absorption of a solute and a solvent (e.g., ^{3}H-labeled water and ^{14}C-labeled solute). A combination of two radioisotopes may be used to check the integrity of the skin barrier. In such experiments flux of tritiated water is measured for several hours and only the skin samples providing an appropriate flux are selected for studying penetration of the compound under investigation, which is ^{14}C labeled, in turn.

Before putting radioisotopes on the skin in a diffusion chamber, principles of handling radiochemicals, which may become crucial for the experiments, should be learned.

I.   STOCK SOLUTIONS AND PREPARATION OF FORMULATIONS FOR PERCUTANEOUS PENETRATION STUDIES

Radioisotopes arrive from a manufacturer in a solid or liquid form specific for a pure substance, or, much more often, in a solution form in solvents such as water, ethanol, benzene, and methanol. For easier dispensing stock solutions are prepared. This is also recommended for many radioisotopes to enhance their stability.[1] It is good to know that addition of 1 to 2% of ethanol, glycerol, or benzyl alcohol to the solution is very beneficial, because these are good free radical scavengers minimizing the self-decomposition of the tracer.[1]

In some cases, particularly in *in vivo* studies, radioisotopes are applied to the skin as they are received from the supplier, but generally dilutions in an appropriate solvent and with an unlabeled compound are prepared. When further dilution of the radioisotope before administration on the skin is not intended or the volume of the stock solution used for preparation of the donor phase is relatively large, attention should be paid if a solvent chosen for a stock solution does not act as a percutaneous penetration enhancer. This may concern most organic solvents.

High specific radioactivities of radioisotopes and high cost are the reasons why in most cases mixing of radioactive ("hot") and nonradioactive ("cold") compounds is required. This procedure is also necessary to establish an effective thermodynamic activity of the penetrant on the skin. To make it clearer, when, for example, the specific radioactivity of [^{3}H]histidine is 50 Ci/mmol and we plan to use 0.1% solution as a donor fluid, to imitate the *in vivo* conditions, then, even if we apply only a small amount of the solution, such as 100 μl on 1 cm^2 of skin, the radioactivity in a single diffusion chamber will be as high as 30 mCi, which is far beyond the accepted ranges (values up to 4 μCi are reasonable, I would say). To avoid such high radioactivity and waste a donor solution using 0.1% "cold" solution and an amount of the "hot" stock solution to obtain required radioactivity should be prepared. The "cold" compound gives the proper concentration gradient necessary for the diffusion process across the skin while the "hot" compound provides a tool to measure the penetrant concentration in the skin or in the receiver fluid.

The act of mixing radioactive and nonradioactive compounds in solutions is called "spiking" and usually is made in such a manner that a specified volume of a stock radioactive solution is added to the solution of "cold" compound just before applying it on the skin. More problematic seems to be the preparation of a radiolabeled saturated solution of the penetrant. In this case mixing the "hot" and "cold" compounds cannot be avoided due to a high amount of the compound required. As stock solutions of radioisotopes are usually used, preparation of the saturation should be done in steps: (1) evaporation of the solvent of the stock solution, (2) combining, in the same vial, the residue with a "cold" compound in substantia, and (3) adding an adequate solvent. Alternatively, a stock "hot" solution may be added to the suspension of the "cold" solution in a solvent, but at the beginning of equilibration since a certain time is necessary to equilibrate not only sediment and solution but also "hot" and "cold" in the both phases. The latter procedure is possible only when addition of the solvent present in a stock solution does not change solubility of the compound in the solvent under investigation. This is the case when both solvents are the same or when the volume ratios are large enough to eliminate the possibility of interference.

When percutaneous penetration of drugs is studied, the drugs are often applied to the skin in the form of gels or ointments. Radioisotope, in a stock solution, may be combined with other components while making a gel or may be added to the finished formulation, on condition of scrupulous mixing. To prepare lipophilic ointment a stock solution of the "hot" compound may be introduced directly to the ointment base, if the solvent is compatible with the vehicle, or it must be evaporated and the residue is mixed with the vehicle. In both cases general rules of ointment compounding must be observed to obtain uniform dispersion: first a drug must be mixed with a small portion of the vehicle and only then the resulting "concentrate" is mixed with the rest of the vehicle.

Measurement of the initial radioactivity of a solution applied to the skin is always recommended, rather than relying on calculated values; this is especially important in the case of saturated solutions and semisolid formulations. Adsorption to the walls of a vial during evaporation, not sufficient equilibration (saturation) time or not careful mixing, may become a source of serious deviation between calculated and measured radioactivity.

II. PURITY OF RADIOCHEMICALS

The tracer technique has the disadvantage of poor specificity. Impurities can often be labeled during synthesis as efficiently as the principal compound and the detector will count the tracer isotope, irrespective of the molecule in which it resides. An investigator should never put too much trust in the purity and identity of radioisotopes and use them uncritically.

Radioisotopes always come with specified rates of degradation and purity. A radiochemical purity of at least 98% is desirable, while chemical purity should match that of the unlabeled compound. A labeled compound always partially decomposes during storage, even though it was radiochemically pure upon arrival from the supplier. The rate of degradation is often specified by a producer. For example, in optimal storage conditions, decomposition of ^{14}C-labeled chemicals is 1 to 3% per annum, while ^{3}H-labeled compounds decompose with a rate 1 to 3% per month. The degradation may be far larger, depending on storage conditions and an actual batch preparation. Although lower temperatures are preferable, freezing of solutions is not accepted.[1] A very fundamental point to note is that the rate of decomposition of a radiochemical may accelerate after an initial period of apparent stability.[2]

Even a small amount of decomposition by-product with high specific activities may disturb the outcome of an experiment. The ideal is always to purify the radioisotope just prior to the experiments, employing thin-layer chromatography (TLC) or other chromatographic techniques. This is time consuming and laborious, however. If our estimate is that this is not necessary, then we should use the radioisotope as soon as possible, check purity regularly, and be very careful while interpreting the results received from a scintillation counter.

Whether radioactive impurities in a tracer interfere and result in artifacts or misleading data depends largely upon the nature and behavior of these impurities in the experimental system. When the compound under investigation penetrates the skin rather easily, the dose diffusing to the receiver fluid is large in comparison with the content of impurities in the radioisotope and the experiment lasts only a few hours, then we may trust that the radioactivity measured in the receiver fluid comes from the studied compound.

The problem, however, starts when the compound penetrates the skin only in small amounts, such as <1% of the dose applied. In such cases even small impurities of 0.1 to 1% make a correct interpretation of the measurements impossible. We never know if the radioactivity in the receiver fluid or in the skin shows penetration of the compound studied or its radioactive impurities. The time of the diffusion experiments ranges from a few hours to 4 days. This is the time of contact of biological tissue with the radioisotope: the longer the time the greater is the chance of degradation of the radioactive and nonradioactive forms of the penetrant. Measurement does not directly give information on the type or form of the product associated with the radioactivity. Thus, despite direct purification of the tracer before the diffusion experiment, one must not uncritically trust the results of the radioactivity measurements in the receiver fluid and in the skin. Even minimal degradation, if it leads to the formation of a more permeable product, when the rate of degradation is higher than the rate of absorption, may result in a high percentage of the degraded compound in the receiver fluid; in some cases the total radioactivity may come from the degraded compound. Radioactivity measurements give an overestimation of the skin penetration of the compound under investigation.

Macromolecules will decompose at a much faster rate than simple molecules and it is especially important to verify the results obtained using radiolabeled compounds. Studies on skin penetration of radioisotopes of collagen, heparin, or insulin must include experiments proving the identity of the radioactive molecules in the receiver fluid and in the skin, otherwise they cannot be reliable.

It must be taken into consideration that residual enzymatic activity of an excised skin or bacterial degradation is a likely source of radioactivity not attributed to the compound under investigation. A radiochemically pure and stable compound does not necessarily exhibit a corresponding stability in a biological system. A labeled molecule may lose its tracer in the course of unrecognized metabolic reactions.

Unexpectedly high fluxes or unusual absorption profiles must be the first signals for checking the radiochemical purity and identity of the penetrant in the receiver fluid and in the skin. Verification of the results with other radionuclides (e.g., ^{3}H vs. ^{14}C), radiochromatography, or other analytical methods should be advised while using radioisotopes.

III. SAMPLE PREPARATION FOR RADIOACTIVITY MEASUREMENT

The most convenient technique for determination of β-emitters ^3H and ^{14}C is liquid scintillation counting. This has its attendant problems, in particular, poor miscibility of scintillation cocktails with water, and high backgrounds. A suitable scintillation cocktail, with good emulsifying properties, should be chosen for measurement of radioactivity in an aqueous receiver fluid. One of the most popular surface agents used to produce emulsions for scintillation counting is Triton X-100. In commercial scintillation liquids this agent is mixed with toluene at a ratio 1:2 (v/v) and the mixture can incorporate as much as 43% water.[1,4] At lower concentrations, usually below 15%, a clear solution, with good counting efficiencies, is formed. As the water content is increased, the emulsion becomes unstable and separates into a two-phase system. This type of system is of little value for counting aqueous samples. However, as the water content is increased even further, up to 20% and more, a stable emulsion forms. This emulsion gives somewhat lower counting efficiencies than the clear system formed at lower water concentrations, so in general the ratios of scintillation cocktail/aqueous sample must be carefully controlled to obtain reproducible results. These ratios appear to be more critical for tritium than for ^{14}C.[4]

When a flow-through diffusion cell is used, usually radioactivity of the whole collected fraction of the receiver fluid is measured. On the basis of my own experience with some scintillation cocktails, I would suggest that one should collect maximum portions of 3 ml, while using 20-ml scintillation vials. When the volume is higher, then separation of phases may occur, if not immediately, then later, when samples are left in a scintillation counter for several hours. The diffusion experiment must be planned taking this into consideration.

When radioactivity of skin samples is to be measured, the tissue must be solubilized. Dimethyldialkyl quaternary ammonium hydroxides, toluene-soluble quaternary ammonium bases, are used to digest biological samples.[4] Among these are commercial products: NCS (Nuclear-Chicago), Soluene (Packard), Bio-Solv (Beckman), Protosol (New England Nuclear), and Eastman Tissue Solubilizer (Eastman Kodak). They permit direct incorporation of aqueous samples into the cocktail. Digestion with these bases requires some attention to detail. The sample to be digested should be covered completely so that it remains in contact with the solubilizing agent; sometimes gentle agitation is helpful. To speed the process, samples may be warmed at a temperature of 37°C. Usually 24 h digestion in these conditions is enough, but it depends on the amount of tissue and its water content. All the bases react with the tinfoil cap lining to produce products that cause considerable quenching.[4] Therefore it is important to use caps lined with polyethylene or Teflon. The preference for one solubilizer over another is simply a matter of experiment. It is suggested that 1 ml of the solubilizer is required for digestion of 50 mg of dry and

20 to 200 mg of wet tissue. To reduce the amount of the solubilizer and minimize background the skin corresponding only to the surface of penetration should be excised and digested.

Colored solutions are frequently formed with quaternary ammonium solubilizers. This color can usually be dissipated by the addition of a solution of 30% H_2O_2 in water or 20% benzoyl peroxide in toluene.[1] Chemilumines-cence can be decreased or eliminated by the addition of a few drops of freshly prepared 15% ascorbic acid in water or a few drops of glacial acetic acid, leaving the samples in the dark for 24 h.

Determination of the compound distribution in stratum corneum layers removed with an adhesive tape is possible by a simple direct measurement. A piece of the tape with adhering horny layers is placed in a vial and scintillation cocktail is added. Care must be taken to avoid the tape adhering to the glass, and intense shaking is recommended. As always, background measurements of the adhesive tape are necessary.

IV. TRITIUM-LABELED PENETRANTS

The least reliable are results obtained with tritium-labeled compounds. Tritium can exchange spontaneously with hydrogen atoms located elsewhere. It is well known that due to proton exchange tritiated water is formed on storage of many ^3H-labeled compounds in aqueous solutions.[3]

It happens that even manufacturers send samples of unstable ^3H radioiso-topes in aqueous solutions, which may cause serious degradation. I faced such a problem when over 90% of the radioactivity in a radioactive peptide just received was attributed to tritiated water. Chromatographic analysis by TLC or paper seldom identifies the effect of tritiated water formation and a direct measurement of tritium activity in a lyophilized sample is usually required.[2] A very simple and fast way to determine if the sample is contaminated with tritiated water is evaporation, under the condition that the compound is not volatile. A sample of the radioactive stock solution may be placed on a filter paper and left for evaporation. The radiation after evaporation must equal the initial value. Tritiated water can be removed by freeze-drying and redissolving the compound immediately before use.

Verification of the data from tritium-labeled compounds with another tracer, namely ^{14}C, in parallel control experiments, is recommended.

For separation of the *in situ* formed tritiated water, solid phase extraction may be successfully employed. I have been using this technique in experiments on percutaneous absorption of amino acids. In those experiments L-[2,3-^3H]aspartic acid and L-[ring-2,5-^3H]histidine with declared radiochemical purity of 99% and L-[4,5-^3H]lysine with purity of 97.8% were used. Radiochemicals were evapo-rated to dryness in a stream of nitrogen, directly prior to the preparation of the experimental solutions and application to the skin. In spite of this procedure, tritiated water was formed in the donor solution during the experiment.

In vitro percutaneous absorption of amino acids was investigated with human dermatomed skin. Amino acid solutions of concentration 0.3 to 1% and pH 3.5 to 8.9, labeled with adequate radioisotopes to obtain a radioactivity of approx. 10 µCi/ml, were placed on the skin (300 µl/cm^2) and served as a donor phase. The receiver solution was 0.9% NaCl. Both solutions, donor and receiver, contained gentamicin sulfate to prevent bacterial growth. Samples of the receiver fluid were collected every 10 or 14 h, up to 86 h. The permeation experiment was described in detail elsewhere.[5]

An ion-exchange column (1 × 7 cm) was prepared using mixed bed resin AGR 501-X8 (D) (Bio-Rad Lab, Richmond, CA). The collected receiver fluid fractions were adjusted to pH 5 with 1 *M* HCl. A volume of 0.5 ml was placed on the column, and the tritiated water was removed from the column with 0.025 *M* phthalate buffer pH 5.0 until no radioactivity was detected in the efflux. Approximately 9 ml of the buffer was used. The total radioactivity of the 0.5 ml sample and the radioactivity washed out from the column were measured; the amount of the radiolabeled amino acid in the sample was calculated as the difference between these two values. Tritiated water (300 to 30,000 dpm) or standard solutions of amino acids (300 to 30,000 dpm) were placed on the column to estimate accuracy and reproducibility of this separation technique. In 10 experiments, when tritiated water was rinsed from an ion-exchange column, 96.4 ± 3.2% (mean ± SD) of the initial radioactivity was recovered in the efflux. When ^{14}C-labeled amino acids were dissolved in buffer solution at pH 5.0 and placed on the column, no radioactivity was detected in the portion washed out with the buffer. To substantiate these measurements, samples of selected receiver and donor solutions were also evaporated to dryness, and the radioactivity in the residue was measured. The mean ratio ($n = 43$) of the radioactivity measured in the residue to that by the ion-exchange column method was 0.99 ± 0.23, but there was a trend for this ratio to increase with the content of tritiated water in the sample.

After 86 h of the diffusion experiment performed at a temperature of 37°C it was found that various amounts of radioactivity were not bound to the ion-exchange column and this portion was assumed to be tritiated water formed *in situ*. The presence of tritiated water in the receiver fluid was related to its content in the donor fluid, which is shown in detail in Table 1. The amount of the tritiated water formed in the donor fluid varied from experiment to experiment, and also depended on the time of its duration. Consequently it was not possible to predict what portion of the radioactivity was attributed to the amino acids and generally the separation had to be performed for all samples of the receiver fluid.

The flux of tritiated water formed *in situ* was, in most cases, 0.6 to 1.3% per hour of its actual content in the donor fluid, what is in agreement with the value 1.05 ± 0.37% of the initial dose per hour measured in independent experiments with tritiated water.

From Table 1 it may be concluded that if the formation of tritiated water had not been discovered the flux of amino acids could have been overestimated

TABLE 1
***In Situ* Formation of Tritiated Water after 86 h of *in Vitro* Skin Permeation by Tritium-Labeled Amino Acids**

Tritiated water in total radioactivity (%)	
Donor fluid	Receiver fluid
0.5	3–10
0.9	20
1.4	25–40
3.1	40–50
10.6	40–65
16.2	55–80

by 10% up to 500%. The most serious was, however, overestimation of lag time. In the first series of experiments, when the tritiated water was not yet determined, 14 h was considered to be enough time to measure steady-state flux since t_{lag} was estimated to be about 1 h. Actually this was t_{lag} of water diffusion across the skin, while the actual t_{lag} for amino acids was 20 to 30 h. These results were also verified using ^{14}C-labeled amino acids.

To remove tritiated water from the skin, samples of the tissue collected after the experiment were dried *in vacuo* above P_2O_5 for at least 2 days. Before digestion the tissue was rehydrated with one or two drops of water to make the solubilization easier.

It is interesting that formation of tritiated water in the course of the diffusion experiment does not result from the contact with biological material as could be expected. In Table 2 tritiated water content in the solutions stored for 48 h at a temperature of 37°C is compared with that in the same solutions when a piece of dermis was soaked in it. No difference is noted. Also pH does

TABLE 2
Formation of Tritiated Water in Solutions of Tritium-Labeled Amino Acids after 48 h of Storage at a Temperature of 37°C in the Presence of Dermis

		Tritiated water in the total radioactivity (%)	
Amino acid	pH	Standard solution	Standard solution + dermis
Aspartic acid	3.4	0.9	1.0
	7.3	1.0	3.0
Histidine	5.0	4.4	4.4
	7.3	14.7	15.0
Lysine	7.3	1.7	1.8
	8.9	1.3	1.3

not seem to play a role in the process of tritium exchange, although generally this process is considered to be influenced by pH.[2] Thus the process is rather temperature dependent, although further studies are required to make such a conclusion trustworthy.

The above example illustrates the need to check for the presence of radiochemical impurities and their effects to interpret the results with confidence.

My advice is that one should rather consider using alternative methods of analysis before turning to radioisotopes. Modern chromatographic techniques are particularly recommended, since not only quantitative determination but also identification of the penetrant is done and eventual metabolism or degradation may be easily discovered. However, the radiochemical method is still invaluable when determination of minute concentrations, below the sensitivity of other methods, is required.

A clear understanding of all the above topics is essential for properly maintaining a tracer compound and correct interpretation of the subsequent experimental data. Let the following quotation be a warning against uncritical conclusions: "It is indeed exceedingly fortunate that many investigations using impure radiochemicals have yielded qualitatively correct results in that the impurities have either failed to incorporate into the biological system or have been removed during the treatment of the sample. One is still left wondering just how many results in the published literature are erroneous in this case."[2]

ACKNOWLEDGMENTS

The author is grateful to Prof. Bartlomiej Kwiatkowski for helpful suggestions during the preparation of the manuscript.

References

1. **Evans, E.A. and Muramatsu, M., Eds.,** *Radiotracer Techniques and Applications*, Vol.1, Marcel Dekker, New York, 1977.
2. **Evans, E.A.,** Purity and stability of radiochemicals, in *Isotopes: Essential Chemistry and Applications*, Elvidge, J.A. and Jones, J.R., Eds., The Chemical Society, London, 1980, 67.
3. **Feinendegen, L.E.,** *Tritium-Labeled Molecules in Biology and Medicine*, Academic Press, New York, 1967.
4. **Munson, J.W., Ed.,** Principles of liquid scintillation counting, in *Pharmaceutical Analysis. Modern Methods*, Marcel Dekker, New York, 1981, 325.
5. **Sznitowska, M., Berner, B., and Maibach, H.I.,** In vitro permeation of human skin by multipolar ions, *Int. J. Pharm.*, 99, 43, 1993.

16 Phototrichogram Analysis of Hair Follicle Stimulation: A Pilot Clinical Study with a Peptide–Copper Complex

Ronald E. Trachy, Leonard M. Patt, Gordon M. Duncan, and Bernard Kalis

TABLE OF CONTENTS

I.   INTRODUCTION

A class of peptide–copper complexes has been described that possesses potent *in vitro* and *in vivo* biological activities.[1-4] These activities include white cell chemoattraction, angiogenesis, and collagen and proteoglycan synthesis. One of these compounds, PC1031 ([glycyl-L-histidyl-L-lysine-L-valine-L-phenylalanine-L-valine]$_2$ - Cu), stimulates hair growth in several animal models. When injected intradermally at the beginning of the long second telogen cycle in mice (typically age 50 to 95 days), PC1031 and other peptide–copper complexes cause the transformation of resting (telogen) follicles into the growing (anagen) phase, with hair growth visible within 10 to 14 days.[5,6] Topical application of PC1031 to mice causes the same phenomenon within 21 days. These results are similar to those shown with topically applied diazoxide[7] and minoxidil.[8] The fuzzy rat[9] has been described by Uno[10] as an effective model for studying topical hair growth compounds such as minoxidil. Using PC1031 in this animal model, Trachy et al.[11,12] demonstrated a positive increase in the proportion of anagen hairs and an increase in hair follicle size (diameter and depth) and length of hair, suggesting a prolongation of the anagen phase.

A quantitative analysis technique known as the phototrichogram[13] provides a noninvasive method for evaluating hair growth in humans. The phototrichogram evaluates shifts in hair follicle populations (anagen and telogen) as well as total hair density. The natural course of male pattern baldness involves a decrease in the proportion of anagen hairs, together with a progressive decrease in the diameter and depth of the hair follicles (miniaturization), resulting in the conversion of terminal (long, large diameter) hair to vellus (fine) hair.[14]

The animal model results with peptide–copper complexes indicates that these compounds can initiate a change in the hair follicle cycle from telogen to anagen and extend the duration of the anagen phase. This ability to extend the length of the anagen cycle might, therefore, have a positive outcome with respect to halting the miniaturization process of hair follicles in male pattern baldness.

II. METHODS

This evaluation of a peptide–copper complex, PC1031, was performed in a clinical study following approval by the institutional review committee at the University of Reims, Hospital Sebastopol. Informed consent was obtained from each subject prior to inclusion in the study.

Of the 18 subjects completing the study, 5 were vehicle-treated, 7 were treated with 2% PC1031, and 6 were treated with 10% PC1031. Subjects were randomly assigned to treatment groups.

A. SUBJECT SELECTION

Twenty male subjects (age 21 to 46) diagnosed as exhibiting male pattern baldness (type III–V Hamilton classification[15]) were enrolled. Entry into the study was restricted to subjects with a measurable hair density of more than

100 hairs/cm^2 within the balding area. Subjects were required to be dark haired, Caucasian males not receiving treatment for their hair loss within the previous year and not receiving treatment for a medical condition from which hair loss or hypertrichosis were known effects. Subjects were required to attend bimonthly follow-up examinations during the 6-month treatment period.

B. TEST ARTICLE AND FORMULATION

PC1031 is a peptide glycyl-L-histidyl-L-lysine-L-valine-L-phenylalanine-L-valine complexed with copper at a 2:1 molar ratio of peptide to copper. Two and ten percent concentrations of PC1031 were formulated in a vehicle solution consisting of propylene glycol USP, ethanol USP, hydroxypropylmethylcellulose USP, sterile water USP, and nonoxynol 9 USP (Igepal® CO-630). PC1031 and vehicle (placebo) were supplied in 50-ml dropper bottles. Application of 0.25 ml was made morning and evening by the subject. The formulation was spread around the balding region by finger tips and gently massaged into the scalp. Shampooing was limited to two to three times per week with a nonionic shampoo.

C. PHOTOTRICHOGRAM TECHNIQUE

Prior to treatment initiation, a baseline phototrichogram was obtained (month 0). Obtaining a phototrichogram was a two clinic visit procedure. On day 0, an approximately 1 cm^2 site located within the balding vertex region (referred to as the reference site) was delineated. Hair was shaved from the area, the reference site tattooed at two diagonal corners, and an accurate square centimeter measurement of the reference area was obtained. A 2× photograph of the reference area (including scale bar) was then taken. The subject returned 3 days later for a second photograph of the shaved area. From these photographs, hair population and comparative growth data were obtained.

Identification of single hairs serving as landmarks allowed the evaluator to generate data with respect to the number of anagen hairs (measurable increase in length from day 0 to 3) and of telogen hairs (no change in length). The number of these two hair types was counted within the controlled reference area. All observed (measurable) hair within the reference area was marked on the photograph including terminal (thick, pigmented hairs), nonvellus (well-defined but thin pigmented hairs), and vellus hairs, which were visibly defined. This procedure (except for tattooing) was followed for subsequent evaluations at months 2, 4, and 6.

The principal investigator and the phototrichogram evaluator were blinded as to treatment group assignments.

TABLE 1
Study Demographics

	Treatment group		
	Vehicle	2% PC1031	10% PC1031
Number	5	7	6
Mean age	33	39	30
Age range	21–45	32–46	25–43
Mean duration (years)	7.4	9.3	7.5

D. DATA ANALYSIS

Bimonthly phototrichograms were used for the evaluation of efficacy. Hair counts per reference area were converted to hairs/cm^2 for comparative analysis. Data were generated with respect to measurable hair density (visible vellus, nonvellus, and terminal hairs counted per cm^2) and anagen hair density (growing hairs counted per cm^2). Group data represent the mean and standard error (SEM) of these parameters for subjects within treatment groups. Changes in hair densities were compared across (ANOVA) and within (Student's paired t-test) treatment groups.

Differences in the initial number of hairs/cm^2 were compensated for by expressing data as a percent change from the baseline value. Group comparisons for significant differences were preformed using a Wilcoxin rank sum test. In the event of missed evaluations (two cases), the previous month's values were carried forward.

III. RESULTS

Eighteen subjects completed the 6-month treatment protocol. One of the original 20 subjects was lost to follow-up, while a second subject elected to drop from the study for cosmetic reasons (formulation related). Neither of these subjects completed the first month of treatment. Baseline demographic profiles of the 18 subjects are presented in Table 1.

Individual and group data from the phototrichogram analysis are given in Table 2. At the outset of the study (month 0), there were no statistically significant differences among the groups with respect to either measurable or anagen hair density.

A. MEASURABLE HAIR DENSITY

Over the course of 6 months, there was a statistically significant loss of measurable hair in the vehicle treatment group (average of 26 hairs; $p < 0.05$).

TABLE 2
Measurable and Anagen Hair Density

Treatment group	Subject No.	Measurable hair density (hairs/cm^2) Month				Anagen hair density (anagen hairs/cm^2) Month			
		0	2	4	6	0	2	4	6
Vehicle	1	107	127	104	97	80	100	89	78
	5	209	212	207	149	127	124	120	100
	9	293	270	270[a]	263	195	202	202[a]	200
	11	109	81	104	119	58	71	73	66
	13	159	133	100	118	104	83	64	81
	Mean	175	165	157	149	113	116	110	105
	SEM	35	34	35	30	24	23	25	24
2% PC1031	3	111	109	121	104	68	60	71	50
	6	158	160	162	154	110	90	106	82
	10	238	196	213	258	118	118	116	171
	15	296	236	213	228	240	194	166	191
	17	141	145	141	139	80	86	82	103
	18	227	175	212	212[a]	173	108	129	129[a]
	19	189	189	191	224	128	107	125	111
	Mean	194	173	179	188	131	109	114	120
	SEM	24	15	14	21	22	16	12	19
10% PC1031	2	153	173	183	191	109	117	117	138
	4	171	182	183	171	129	124	99	109
	8	135	163	157	153	55	93	102	95
	12	207	247	193	193	76	159	137	133
	16	170	179	191	182	107	126	139	127
	20	202	170	179	198	106	98	135	164
	Mean	173	186	181	181	97	120	122	128
	SEM	11	13	5	7	11	10	7	10

[a] Missing data, last value carried forward.

When expressed as a percentage of baseline values, there was a significant ($p < 0.05$) difference at 6 months between the mean percentage increase in measurable hair density experienced by the 10% PC1031 treatment group (6.04 ± 4.76% SEM) when compared to the decrease measured in the vehicle treatment group (–13.13 ± 6.85% SEM, Figure 1). The 2% PC1031 treatment group had no change in measurable hair density at 6 months. Four of the five vehicle-treated subjects experienced a decrease in measurable hair density over the course of the study, while three of the six 10% PC1031-treated subjects had an increase in density.

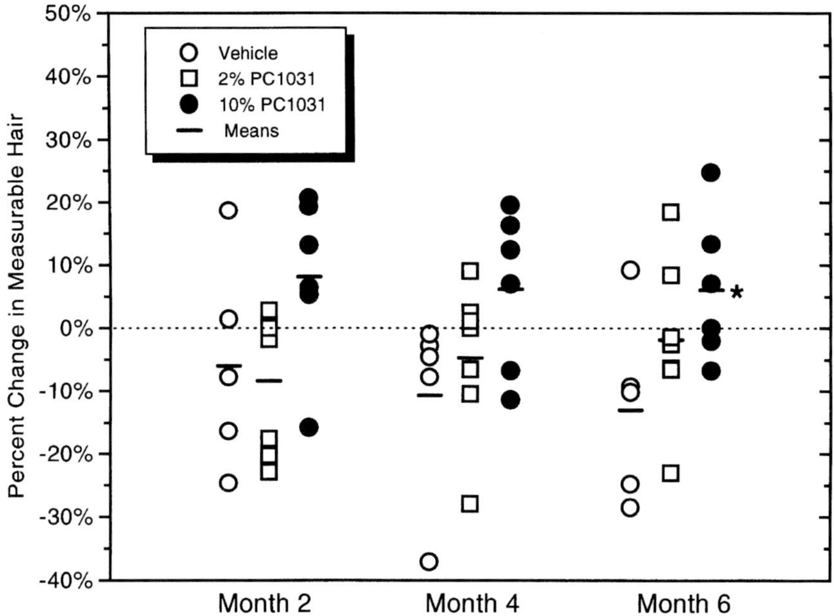

FIGURE 1   Percent change in measurable hair density. Data shown are the percent change from baseline for individual subjects treated with vehicle (open circles), 2% PC1031 (open squares), and 10% PC1031 (closed circles). Subjects in the vehicle-treated group experienced a statistically significant loss of hair over the 6-month study (paired t test, $p < 0.05$), while subjects in the 10% PC1031 group remained relatively stable. At the 6-month time point, a statistically significant difference (*) was found between the vehicle and the 10% PC1031 group (Wilcoxin rank sum text, $p < 0.05$).

B.   ANAGEN HAIR DENSITY

Ten percent PC1031 significantly ($p < 0.05$) increased anagen hair density during the 6-month study, with five of the six subjects showing a positive response (mean group increase of approximately 30 anagen hairs) relative to their baseline values. Anagen hair density did not change significantly following treatment with either the vehicle (three of five subjects decreased) or 2% PC1031 (five of seven decreased). Compared to baseline, anagen hair density increased 39% in the 10% PC1031-treated subjects. This was significant ($p < 0.05$) when compared to the 6% decrease experienced by the vehicle group (Figure 2). The 2% PC1031 group had a 5% decrease in anagen hair density during the same 6-month period.

C.   SUBJECTIVE ASSESSMENT

When questioned regarding the perceived benefit of 6 months of treatment (reduced loss of hair, increase in hair density), three of the five vehicle-treated

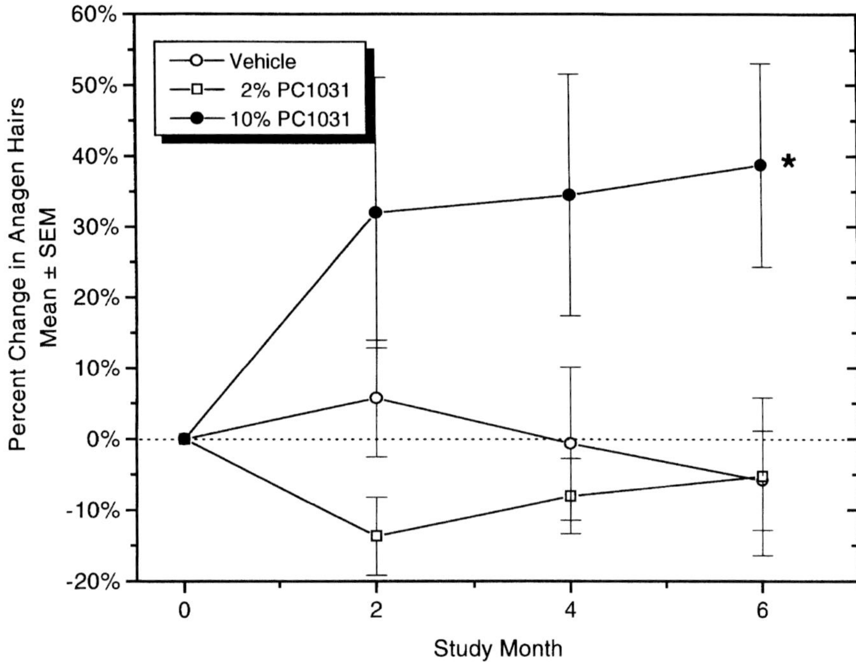

FIGURE 2 Mean percent change in anagen hair density. Data shown are the mean ± SEM of the percent change in the anagen hair density from baseline calculated from the data in Table 2. There was a statistically significant different (*) between the 10% PC1031 and the vehicle group at the 6-month time point (Wilcoxin rank sum test, p <0.05).

subjects, five of seven 2% PC1031-treated subjects, and four of six 10% PC1031-treated subjects felt they showed improvement in their hair loss condition.

IV. DISCUSSION OF FINDINGS

Treatment of hair loss remains a paramount concern to both patient and clinician. Animal model studies have shown that peptide–copper complexes, in this case, PC1031, are potent stimulators of the anagen phase of the hair follicle. As a continuation of the development of these complexes as potential hair growth agents, PC1031 was tested in male subjects experiencing Stage V androgenetic alopecia. A phototrichogram analysis was used to establish whether this compound could cause a shift in hair follicle cycle distribution in humans similar to that seen in experimental animal studies.

In this blinded, randomized clinical study in men experiencing progressive hair loss (male pattern baldness, androgenetic alopecia), the peptide–copper complex PC1031 was found to stabilize hair loss. At the same time, subjects receiving a vehicle without drug experienced a significant loss of hair over a period of 6 months as would be expected in androgenetic alopecia. A significant increase in anagen hair density was also measured in subjects treated with 10% PC1031 (+39%) while this hair cycle population was not significantly influenced by either 2% PC1031 or vehicle treatment. These findings are consistent with animal model results with PC1031. There were no treatment-related side effects observed.

Although the specific mechanism of action is unknown, PC1031 has been shown to stimulate resting telogen follicles, causing them to become actively growing anagen follicles. Topical application of PC1031 to fuzzy rats over a 5-month time period caused an increase in the anagen population and extension of the growth period, which in turn resulted in the growth of larger follicles and longer hairs.[11,12] These phenomena, increasing the ratio of anagen to telogen hairs and extending the length of the anagen cycle could, over time, result in halting the natural course of male pattern baldness in men (a decrease in the proportion of anagen to telogen hairs), and prevent or delay the miniaturization (the progressive decrease in the hair size and length) of the hair. A longer treatment protocol would need to be utilized to demonstrate this potential.

Comparisons with the minoxidil (Rogaine™) clinical experience are interesting. Evaluation of the clinical effects of minoxidil have utilized several quantitative and qualitative methodologies. In 1985, DeVillez[16] reported that 2 or 3% minoxidil achieved an excellent regrowth effect in 14% of treated subjects. In several clinical trials,[17-22] visual hair counting has provided evidence of hair growth with minoxidil as well as its vehicle. In 1989, Rushton et al.[23] reported that minoxidil failed to stimulate hair regrowth when evaluated using a hair plucking method known as the unit area trichogram. They could not detect a significant increase in hair density after 6 or 12 months treatment with 2% minoxidil. A trend toward an increase in both total hair density and anagen hair density was detected in minoxidil-treated subjects, while a decrease was observed in these parameters in vehicle counterparts. James and Rushton[24] demonstrated problems with the visual hair counting method used extensively in topical minoxidil studies.

The hair densities in the present study were more consistent with the unit area trichogram data (approximately 150 to 300 hairs/cm^2) than with studies using direct hair counting methodologies. Similar to the unit area trichogram results with minoxidil, the phototrichogram results with 10% PC1031 demonstrated an overall trend toward hair regrowth, while the vehicle group experienced a decrease in hair density. The relative efficacy of PC1031 and minoxidil is difficult to assess at this time. However, when evaluated in separate studies utilizing sensitive analysis techniques rather than direct counting, both drugs appear to at least arrest hair loss, and perhaps stimulate hair regrowth.

References

1. **Pickart, L. and Lovejoy, S.,** Biological activity of human plasma copper-binding growth factor glycyl-L-histidyl-L-lysine, *Methods Enzymol.*, 147, 314, 1987.
2. **Zetter, B.R., Rasmussen, N., and Brown, L.,** An in vivo assay for chemoattractant activity, *Lab. Invest.*, 53(3), 362, 1985.
3. **Raju, K.S., Alessandri, G., Ziche, M., and Gullino, P.M.,** Ceruloplasmin, copper ions, and angiogenesis, *JNCI*, 69(5), 1183, 1982.
4. **Maquart, F.-X., Bellon, G., Chaqour, B., Wegrowski, J., Patt, L.M., Trachy, R.E., Monboisse, J.-C., Chastang, F., Birembaut, P., Gillery, P., and Borel, J.-P.,** In vivo stimulation of connective tissue accumulation by the tripeptide copper complex glycyl-L-histidyl-L-lysine-Cu2+ in rat experimental wounds, *J. Clin. Invest.*, 92, 2368, 1993.
5. **Trachy, R.E., Fors, T.D., Pickart, L., and Uno, H.,** The hair follicle stimulating properties of peptide copper complexes: results in C3H mice, *Ann. N.Y. Acad. Sci.*, 642, 468, 1991.
6. **Trachy, R.E., Timpe, E.D., Dunwiddie, I., and Patt, L.M.,** Evaluation of telogen hair follicle stimulation using an *in vivo* model: results with peptide–copper complexes, this volume, chapter 18.
7. **Uno, H., Schroeder, B., Fors, T., and Mori, O.,** Macaque and rodent models for the screening of drugs for stimulating hair growth, *J. Cut. Aging Cosmet. Dermatol.*, 1, 193, 1990.
8. **Takeshita, K., Yamagishi, I., Sugimoto, T., Otomo, S., and Moriwaki, K.,** The hair growing effect of minoxidil, *Ann. N.Y. Acad. Sci.*, 642, 470, 1991.
9. **Ferguson, F.G., Irving, G.W., and Stedham, M.A.,** Three variations of hairlessness associated with albinism in the laboratory rat, *Lab. Anim. Sci.*, 29, 459, 1979.
10. **Uno, H.,** Quantitative models for the study of hair growth, *Ann. N.Y. Acad. Sci.*, 642, 107, 1991.
11. **Trachy, R.E., Packard S., Uno, H., and Pickart, L.,** The fuzzy rat, a model of iatrogenic hair growth, and the effect of PC1031, *J. Invest. Dermatol*, 96, 579, 1991.
12. **Trachy, R.E., Uno, H., Packard, S., and Patt, L.M.,** Quantitative assessment of peptide–copper complex induced hair follicle stimulation using the fuzzy rat, this volume, chapter 17.
13. **Guarrera, M. and Ciulla, M.P.,** Quantitative evaluation of hair loss: the phototrichogram, *J. Appl. Cosmetol.*, 4, 61, 1986.
14. **Rushton, D.H., Ramsay, I.D., Norris, M.J., and Gilkes, J.J.H.,** Natural progression of male pattern baldness in young men, *Clin. Exp. Dermatol.*, 16, 188, 1991.
15. **Hamilton, J.B.,** Patterned loss of hair in man: types and incidence, *Ann. N.Y. Acad. Sci.*, 53, 708, 1951.

16. **DeVillez, R.L.,** Topical minoxidil therapy in hereditary androgenetic alopecia, *Arch. Dermatol.*, 121, 197, 1985.

17. **Olsen, E.A., Weiner, M.S., DeLong, E.R., and Pinnell, S.R.,** Topical minoxidil in early male pattern baldness, *J. Am. Acad. Dermatol.*, 13, 185, 1985.

18. **Civatte, J.,** The European topical minoxidil study group. Topical 2% minoxidil solution in male pattern alopecia: the initial European experience, *Int. J. Dermatol.*, 27, 424, 1988.

19. **Kreindler, T.G.,** Topical minoxidil in early androgenetic alopecia, *J. Am. Acad. Dermatol.*, 16, 718, 1987.

20. **Roberts, J.L.,** Androgenetic alopecia: treatment results with topical minoxidil, *J. Am. Acad. Dermatol.*, 16, 705, 1987.

21. **DeVillez, R.L.,** Androgenetic alopecia treated with topical minoxidil, *J. Am. Acad. Dermatol.*, 16, 669, 1987.

22. **Olsen, E.A., DeLong, E.R., and Weiner, M.S.,** Long-term follow-up of men with male pattern baldness treated with topical minoxidil, *J. Am. Acad. Dermatol.*, 16, 688, 1987.

23. **Rushton, D.H., Unger, W.P., Cotterill, P.C., Kingsley, P., and James, K.C.,** Quantitative assessment of 2% minoxidil in the treatment of male pattern baldness, *Clin. Exp. Dermatol.*, 14, 40, 1989.

24. **James, K.C. and Rushton, D.H.,** Evaluation technics for male pattern baldness, *J. Am. Acad. Dermatol.*, 14, 849, 1986.

17 Quantitative Assessment of Peptide–Copper Complex-Induced Hair Follicle Stimulation Using the Fuzzy Rat

Ronald E. Trachy, Hideo Uno, Shelley Packard, and Leonard M. Patt

TABLE OF CONTENTS

I.   INTRODUCTION

Hair growth is governed in part by the cycling of hair follicles from anagen (growth phase) to catagen (transition phase) to telogen (resting phase). These phases produce cyclic patterns of hair growth in animals and man.[1] The length and diameter of the hair shaft are determined by the length of time the hair follicle remains in the anagen phase. Within the anagen phase follicle population there is an additional variation in the type of hair that is produced, depending on the duration of the anagen phase. Large anagen phase hair follicles (those which have been in anagen phase for a long time) tend to produce what is referred to as terminal hairs (long, relatively large-diameter hair shafts) while smaller anagen follicles, which are progressing through the growth cycle rapidly, produce short, thin, vellus hairs.

The loss of hair is typically associated with clinical conditions such as androgenetic alopecia (also known as male pattern baldness), alopecia areata, female pattern baldness, or iatrogenically induced hypotrichosis (such as che-motherapy- or radiation-induced). In androgenetic alopecia, a progressive miniaturization of terminal hairs occurs as a function of a decrease in the anagen period (normally 3 to 5 years), and a decrease in the proportion of anagen hairs to telogen hairs. The predominance of telogen hairs and the miniaturization of terminal hairs (a decrease in hair follicle size) contribute to a condition of general balding in the affected area of the scalp.[2]

A reversal of this tendency toward progressively shorter and thinner hair may be a function of stimulation of resting (telogen) hair follicles and a prolongation of the active growing (anagen) follicle cycle. The percent of hair follicles in the anagen phase and their relative size are important physiological endpoints that need to be studied with respect to potential hair growth stimulating drugs and treatments.

The fuzzy rat, an animal with hair follicle characteristics that closely mimic human androgenetic alopecia, allows for investigation of hair follicle cycling and growth.

II. DESCRIPTION OF THE TEST SYSTEM

A genetic cross between hairless and haired albino rats,[3] the male fuzzy rat provides a model for studying the effects of potential hair growth-promoting drugs on hair growth and, most importantly, on the cycling of hair follicles. The fuzzy rat has been found to display sex-dependent characteristics with respect to hair growth and hair loss. The characteristic features of the sparse fuzzy hair coating of this genetic mutant rat develop after 2 months of age. During the juvenile age (22 to 40 days), the body is densely covered with fine white hairs. As the animals grow older, around 60 days, the skin surface of male rats develops a brown scaly appearance, while female rats retain their whitish skin similar to that of the juvenile stage. This sexual difference in skin appearance is largely due to hyperplastic sebaceous glands caused by a postpubertal elevation of androgens in the male rats. Hairs in the adult male rat become markedly sparse and are retained mainly as short *fuzzy* hairs. These are similar in appearance to vellus hairs in humans.

The cyclic growth of hair follicles from neonatal to around 50 days of age exhibits a synchronized growth cycle similar to that of normal haired albino rats. The first cycle begins in the neonate and the fully developed anagen phase lasts from 10 to 20 days. The typical fuzzy coat of the male rat appears after the third cycle of the follicles through the anagen phase. Female fuzzy rats maintain a hair pattern of short fuzz, while the backs of adult males produce sporadic long, kinky hairs (1.5 to 2.0 cm in length) and short fuzz (0.2 to 0.4 mm in length). The hair follicles of postadolescent males show no synchronized pattern of cyclic growth and the majority of hair follicles present are of small vellus size (0.7 mm in length) accompanied by a few large follicles that are producing terminal kinky hair shafts (corresponding follicles are approximately 2 mm in depth into the dermis).

By the age of 50 to 60 days, the male fuzzy rat also shows a marked diminution in the number of large terminal hair follicles, a condition similar to that observed in both the human and stumptailed macaque androgenetic alopecia.[4] The few terminal follicles in the adult male appear shrunken when compared to those of younger animals, again similar to those observed in the human condition. It is possible to quantify drug effects on hair growth in the fuzzy rat model by measuring both the gross changes in terminal hair population as well as micromorphological assessment of the follicle cycle distribution.

Typical studies in this model involve topical treatment of test article(s) to groups of male fuzzy rats for a predetermined period, usually 3 to 6 months. End points are changes in hair growth documented by photographs of the treatment areas, and an assessment of hair follicle cycle distribution by analysis of biopsies taken at the treatment site. A useful tool to aid in this assessment is the folliculogram.

A. ANIMALS

Male fuzzy rats for these studies were obtained from Harlan Sprague-Dawley, Inc., Indianapolis, IN. Hair growth studies are best initiated when the rats are 28- to 30-days old.

B. TOPICAL TREATMENTS

In these studies, treatments were performed 5 days per week (Monday to Friday) with topically formulated test compounds. Controls consisted of groups of fuzzy rats treated in a similar manner with the vehicle. Drug or vehicle was applied to the animals lower back and gently massaged into the skin. The skin is usually not washed prior to application of the next dose of test solution.

C. PHOTOGRAPHS

Photographs were taken of individual animals prestudy and at set points throughout the study period. Macrophotography is used to obtain side view photographs illustrating the presence of *fuzzy* hairs and long terminal hairs. These serve as qualitative assessments of treatment effects.

D. FOLLICULOGRAM ANALYSIS OF HAIR FOLLICLES

Hair follicle size and cycle (anagen, catagen, and telogen) were determined by means of a micromorphometric analysis known as a folliculogram.[5,6] This is a three-dimensional reconstruction of all the hair follicles in a given skin biopsy. The folliculogram is a histogram representing the proportion of the follicle population in each of these different phases of growth and their lengths.

Tissue was obtained from the center of the treatment using an 8-mm punch biopsy tool. Serial sections were cut (10 µm) and each section stained with hematoxylin and eosin (H&E). Each stained section was projected onto a piece of paper and the hair follicles traced. These drawings were used to determine the number of anagen, catagen, and telogen follicles, as well as the depth of the follicle by measurements from the skin surface to the base of the follicle. Hair follicles present in each individual section were classified (mid-anagen, A3, or late anagen, A5), catagen, or telogen. All serial sections were examined in sequence to develop a distribution of hair follicle states within the entire 8-mm biopsy sample. In addition to the cycle classification of the follicles, their length (from the skin surface to the base of the follicle) was also measured. Terminal follicles (those capable of producing terminal hair shafts) were differentiated from vellus (those producing vellus *fuzz*) follicles by grouping according to the following:

Telogen follicles	<0.5 mm = vellus
	≥0.5 mm = terminal
Anagen follicles	<1.0 mm = vellus
	≥1.0 mm = terminal

E. CROSS-SECTIONAL ANALYSIS OF HAIR FOLLICLES

An alternative method of hair follicle distribution analysis was also used to demonstrate a shift in the distribution of anagen and telogen phase follicles. In this method, biopsies are obtained in a manner similar to that described above for the folliculogram. After fixation, the biopsies are sectioned horizontally rather than vertically. Each serial section is examined for follicles and the relative cross-sectional area is measured using either tracing after projection onto a flat surface or computer-aided planimetry. Histological sections were evaluated at a standard depth (just below the sebaceous gland). Histologic slides from H&E stained sections were projected and outlined on paper and measured for cross-sectional area using a computerized graphics tablet. A range of follicular cross-sectional areas (up to >1000 μm^2) was used to express the follicular population size distribution.

III. PEPTIDE–COPPER COMPLEXES FOR HAIR GROWTH

The peptide–copper complexes used in these studies come from a family of compounds described in this volume.[15] The parent compound, glycyl-L-histidyl-L-lysine (GHK) was first described as a tripeptide isolated from human plasma that enhanced the growth and viability of a variety of cultured cells.[7-10] When complexed to copper, this compound stimulates a number of interesting biochemical events, including the synthesis of various components of the extracellular matrix. The extracellular matrix has been implicated in the regulation of hair growth.[11] Many analogs of the GHK–Cu structure, including those shown to be active in hair growth models described here and in Chapter 18, are active in stimulating the synthesis of extracellular matrix components.[12,13] Early observations revealed that certain analogs of the GHK peptide structure, when complexed to copper, were potent stimulators of hair growth after either intradermal injection or topical application.[14,15]

A model peptide–copper compound, GHKVFV–Cu, has been tested in the fuzzy rat to determine its activity with respect to hair follicle cycle, and production of terminal hair shafts. This same compound was tested in a pilot clinical study,[16] causing a significant increase in the percent of hairs in anagen phase as well as a stabilization of hair loss over a 6-month treatment period.

The material used in these studies was the peptide–copper complex, GHKVFV–Cu, code name PC1031, formulated for topical application in a vehicle composed of propylene glycol, ethanol, water, and a nonionic surfactant.

IV. TEST RESULTS AND PRESENTATION DATA

The utility of the fuzzy rat model is demonstrated in two studies examining the effect of PC1031. These studies serve to illustrate both the methodology and types of data generated in this model.

A. THE EFFECT OF GHKVFV–CU ON HAIR FOLLICLE CYCLE

Thirty-six male fuzzy rats were randomly divided into 3 groups of 12 animals each and were housed 3 per cage in a temperature- and humidity-controlled environment. At the time of topical treatment initiation, the animals were approximately 28- to 30-days-old. Topical treatments consisted of 0.2 ml of either 2% PC1031, 10% PC1031, or the vehicle solution. Treatments were performed 5 days per week for a total of 5 months. In addition to macrophotographic documentation, folliculograms were generated from tissue biopsies obtained from two to four animals per treatment group at 3 and 5 months as previously described. For comparison of time and treatment group, averaged folliculograms are provided. Each folliculogram is quantified with respect to the size and proportional distribution of telogen, mid-anagen (A3), late-anagen (A5), and catagen hair follicles. Terminal and vellus follicles were differentiated as previously described for telogen and anagen follicles (A5). Mid-phase follicles (catagen and A3) were excluded in this analysis due to their transitional nature.

After 3 months of treatment with either 2% PC1031 or 10% PC1031, increases in the number of long, kinky hairs, and in the length of the corresponding hair follicles was evident in the treated animals when compared to vehicle-treated animals at the same time point. Representative photographs and photomicrographs of a vehicle-treated and 10% PC1031-treated animal are shown in Figure 1A–F. These photographs demonstrate the short vellus fuzz present on vehicle-treated animals and the corresponding histologic section showing a predominance of vellus size follicles (A and B) and the numerous long terminal hairs and corresponding deep terminal hair follicles present on the 2% and 10% PC1031-treated animals (C through F).

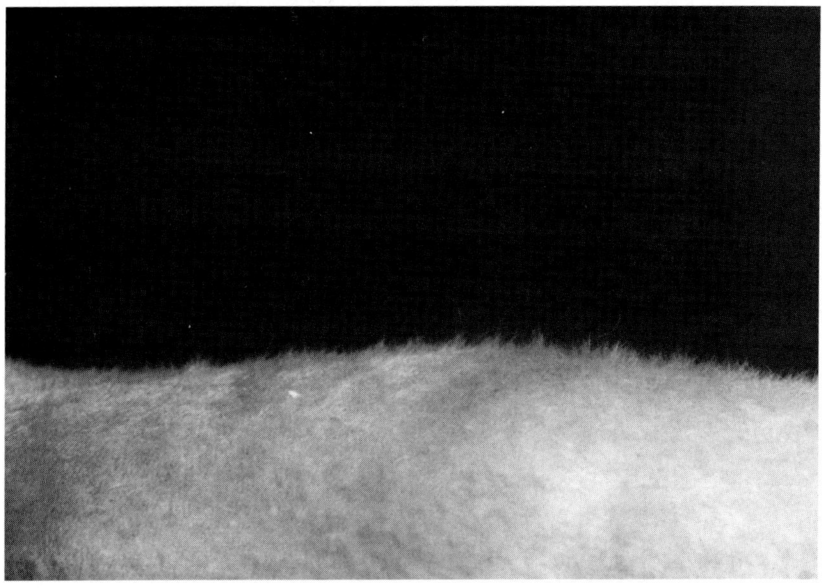

FIGURE 1A–F   Representative photographs and histopathology of fuzzy rats treated with either vehicle, 2% PC1031, or 10% PC1031 at the 3-month treatment point. (A,B) Vehicle-treated animals; (C,D) 2% PC1031 group; (E,F) 10% PC1031 group.

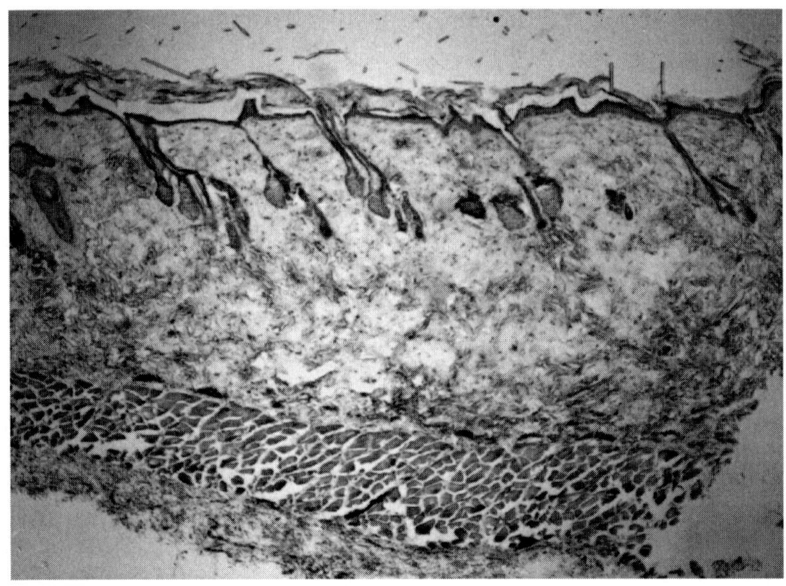

FIGURE 1B

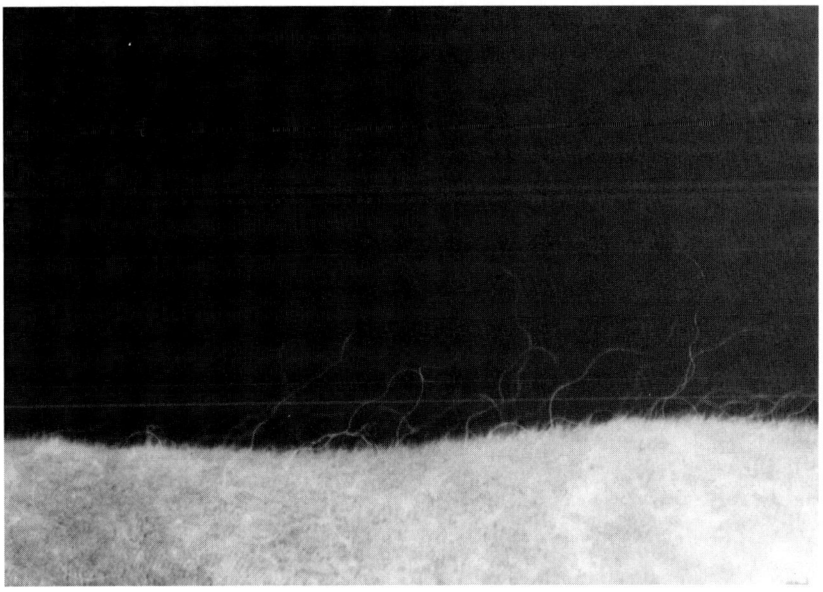

FIGURE 1C

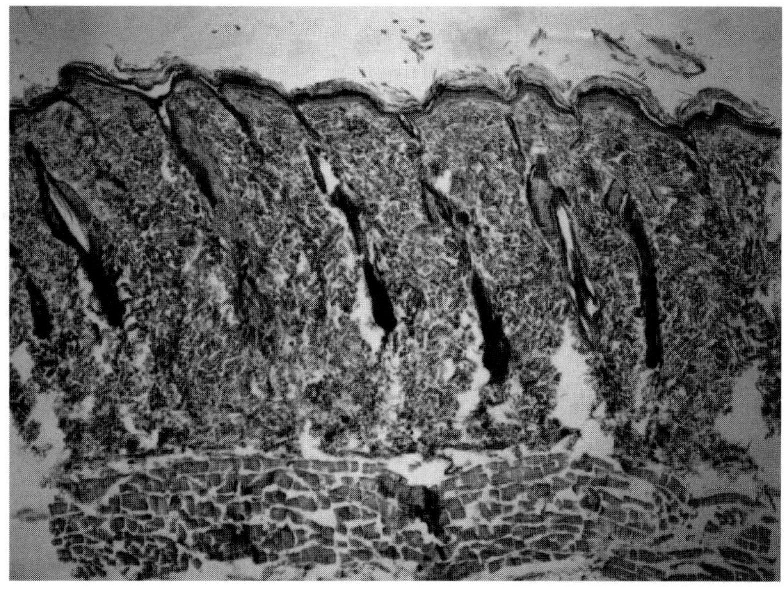

FIGURE 1D

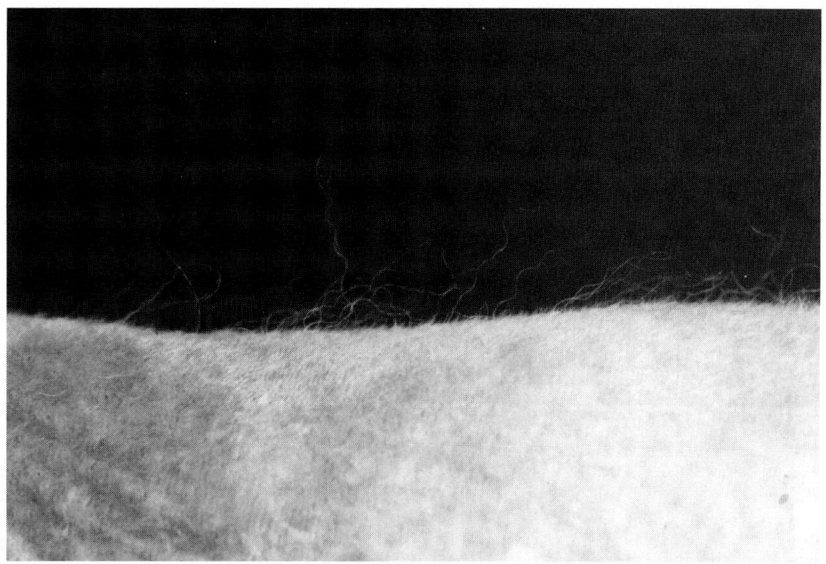

FIGURE 1E

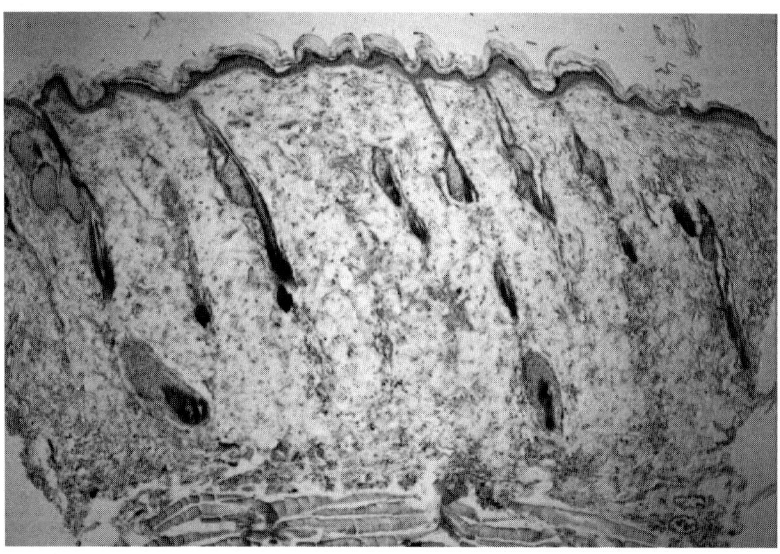

FIGURE 1F

Averaged folliculograms from individual animals are shown in Figure 2. Differences between the PC1031 treatment groups and the vehicle control group were measured with respect to percentage anagen follicle population (A3 + A5) and the percentage of follicles of terminal size. After 3 months of treatment, averaged folliculograms from the two PC1031 groups (2 and 10%) revealed 65 to 70% anagen phase follicles (A3 + A5) as compared to 41% for the vehicle group. The percentage of total follicles in the catagen phase (transition phase between telogen and anagen) was greater in the vehicle group, approximately 37%, compared to 12 to 18% in the PC1031 treatment groups. The percentage of total follicles in the telogen phase was approximately equal in all three treatment groups. By 5 months, the distribution of follicle phase was approximately equal in all three treatment groups.

In addition to evaluation of the follicle phase, the folliculogram analysis allows for measurement of hair follicle depth (an indicator of the length of the corresponding hair shaft). Terminal and vellus size follicles were identified on the individual folliculograms and the percentage of terminal length follicles in each growth phase was calculated. This analysis of follicle size is also presented in Figure 2.

At the 3-month treatment point, approximately 50 to 70% of the anagen follicles and 50 to 90% of the telogen phase follicles were of terminal length (greater than 1.0 mm in depth for the anagen follicles and greater than 0.5 mm in depth for the telogen) following treatment with PC1031. A much smaller

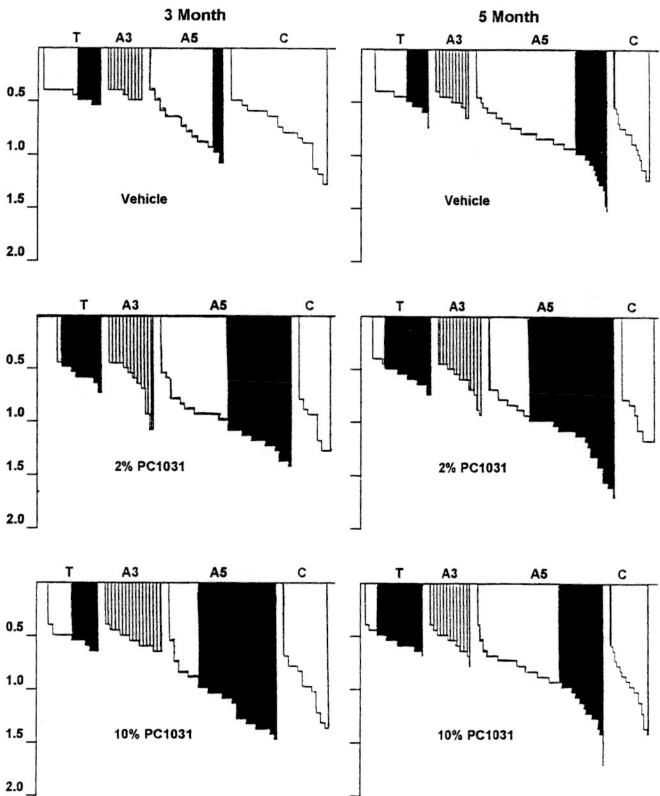

FIGURE 2   Average folliculograms made from histomorphometric analysis of biopsies from fuzzy rats treated with vehicle (top), 2% PC1031 (middle), and 10% PC1031 (bottom). Data from both the 3-month (left side) and 5-month (right side) analysis is shown. Values along the y-axis represent follicle depth (0 to 2 mm), and shaded portion of telogen and anagen 5 represent terminal size follicles.

percentage of the follicles in the vehicle treated group was of terminal length (approximately 15 and 37% respectively). As shown in Figure 2, this trend continued at the 5 month time point.

B.   THE EFFECT OF GHKVFV–CU ON FOLLICLE SIZE (THICKNESS)

Another method to examine changes in follicle size is to determine the cross-sectional area at a constant point. In this aspect of the study, 20 male fuzzy rats were randomly divided into two groups of 10 animals each. Topical treatments consisted of 0.2 ml of either 5% PC1031 or the vehicle formulation. Treatments were performed 5 days per week for a total of 5 months. Tissue biopsies were obtained from the treated area, the follicular cross-sectional

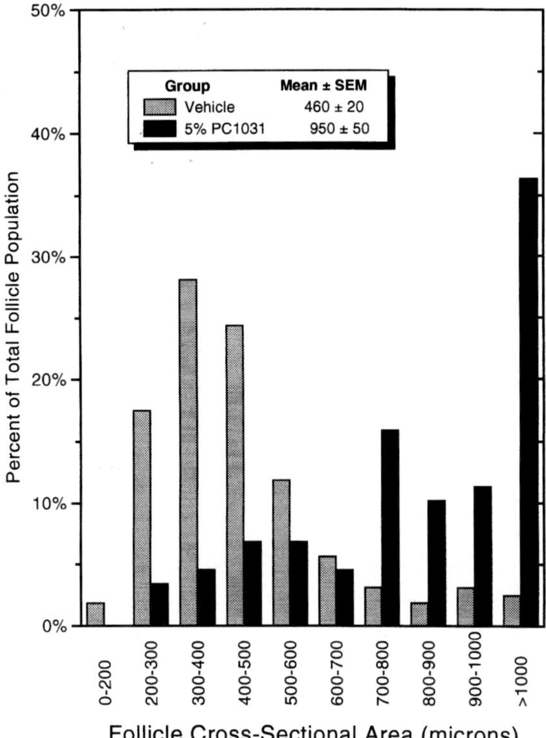

FIGURE 3   Distribution of follicle relative to cross-sectional area. Data shown are the percent of total hair follicles in each treatment group, which are in each cross-sectional area slice.

areas determined, and the follicular population size distribution calculated as a percentage of total follicles in each area category.

Treatment with 5% PC1031 for 5 months resulted in a shift in the follicle cross-sectional area distribution compared to the corresponding vehicle control (Figure 3). The majority of vehicle-treated follicles fell in the range of 100 to 500 μm^2 (mean of 460 ± 20), compared to the majority of PC1031 follicles, which fell into a relative area range of 700 to > 1000 μm^2 (mean of 950 ± 50). This is in agreement with the folliculogram data summarized in Figure 2. Photographic analysis of individual animals in this study also indicated an increased number of the typical long hairs representative of terminal, anagen phase follicles.

V.   INTERPRETATION OF THE MODEL RESULTS

The growth state and type of hair shaft produced by hair follicles are important parameters in the study of potential hair growth-promoting substances and treatments. The utility of the fuzzy rat model to the study of hair growth-promoting compounds is that is allows clear determinations of the relative numbers and sizes of hair follicles in the anagen and telogen phases. This model was used to evaluate the effects of a peptide–copper compound, PC1031, on hair growth. Two general conclusions can be made from the experimental results.

1. Treatment with PC1031 resulted in an increase in the percentage of hair follicles in the anagen or growth phase.
2. Treatment with PC1031 caused an increase in hair follicle size, both in terms of the percentage of telogen and anagen follicles of terminal length and follicle cross-sectional area.

A possible explanation for the increase in follicle size seen in the PC1031 treatment groups is the prolongation of the anagen phase. At 3 months, PC1031-treated animals displayed a greater proportion of anagen follicles when compared to the vehicle-treated animals. A much larger population of catagen follicles was observed in the vehicle group at that time point. An extension of the follicle growth phase may have allowed the shorter hairs to grow to the length of the long *kinky* hairs. If hair follicles are being induced to remain in the anagen state for an extended period of time, this could result in a shift in the proportion of telogen and anagen follicles as well as an increase in the depth of follicles resulting in an increase in the population of long, *kinky* hairs observed in the photographs of the PC1031-treated animals.

These findings suggest that PC1031, a model peptide–copper compound for hair growth, may have significant biological effects on the hair follicle, resulting in both an increase in the percentage of follicles in the anagen phase and prolongation of the anagen phase. These effects on anagen phase result in increasing the length of already existing terminal hairs and in stimulation of the growth of already regressing shorter vellus hairs. These findings also demonstrate the utility and sensitivity of the fuzzy rat model to detect and quantify these changes.

References

1. **Ebling, F.J.,** Hair, *J. Invest. Dermatol.,* 67, 98, 1976.
2. **Rushton, D.H., Ramsay, I.D., Norris, M.J., and Gilkes, J.J.H.,** Natural progression of male pattern baldness in young men, *Clin. Exp. Dermatol.,* 16, 188, 1991.

3. **Ferguson, F.G., Irving, G.W. III, and Stedman, M.A.,** Three variations of hairlessness associated with albinism in the laboratory rat, *Lab. Anim. Sci.*, 29(4), 459, 1979.

4. **Uno, H., Allegra, F., Adachi, K., and Montagna, W.,** Studies of common baldness of the stumptailed macaque (Macaca speciosa). I. Distribution of the hair follicles, *J. Invest. Dermatol.*, 49, 288, 1967.

5. **Uno, H.,** The stumptailed macaque as a model for baldness: effects of minoxidil, *Int. J. Cosmet. Sci.*, 8, 63, 1986.

6. **Uno, H.,** Quantitative models for the study of hair growth in vivo, *Ann. N.Y. Acad. Sci.*, 642, 107, 1991.

7. **Pickart, L. and Thaler, M.,** Tripeptide in human serum which prolongs survival of normal liver cells and stimulates the growth of hepatoma cells, *Nature New Biol.*, 243, 85, 1973.

8. **Schlesinger, D.H., Pickart, L., and Thaler, M.M.,** Growth modulating tripeptide is H-Gly-His-Lys-OH, *Experientia*, 33, 324, 1977.

9. **Pickart, L., Thayer, L., and Thaler, M.,** A synthetic tripeptide which increases survival of normal liver cells and stimulates the growth of hepatoma cells, *Biochem. Biophys. Res. Commun.*, 54, 562, 1973.

10. **Pickart, L.,** The use of glyclyhistidyllysine in culture systems, *In Vitro*, 17, 459, 1981.

11. **Messenger, A.G.,** Extracellular matrix and the hair growth cycle, *J. Invest. Dermatol.*, 75s, 1991.

12. **Maquart, F.-X., Bellon, G., Chaqour, B., Wegrowski, J., Patt, L.M., Trachy, R.E., Monboisse, J.-C., Chastang, F., Birembaut, P., Gillery, P., and Borel, J.-P.,** In vivo stimulation of connective tissue accumulation by the tripeptide copper complex glycyl-L-histidyl-L-lysine-Cu2+ in rat experimental wounds, *J. Clin. Invest.*, 92, 2368, 1993.

13. **Wegrowski, Y., Maquart, F.X., and Borel, J.P.,** Stimulation of sulfated glycosaminoglycan synthesis by the tripeptide-copper complex glycyl-L-histidyl-L-lysine-Cu 2+, *Life Sci.*, 51(13), 1049, 1992.

14. **Trachy, R.E., Fors, T.D., Pickart, L., and Uno, H.,** The hair follicle stimulating properties of peptide:copper complexes: results in C3H mice, *Ann. N.Y. Acad. Sci.*, 642, 468, 1991.

15. **Trachy, R.E., Timpe, E.D., Dunwiddie, I., and Patt, L.M.,** Evaluation of telogen hair follicle stimulation using an *in vivo* model: results with peptide-copper complexes, this volume, chapter 18.

16. **Trachy, R.E., Patt, L.M., Duncan, G.M., and Kalis, B.,** Phototrichogram analysis of hair follicle stimulation: a pilot clinical study with a peptide–copper complex, this volume, chapter 16.

18 Evaluation of Telogen Hair Follicle Stimulation Using an *In Vivo* Model: Results with Peptide–Copper Complexes

Ronald E. Trachy, Erika D. Timpe, Irene Dunwiddie, and Leonard M. Patt

TABLE OF CONTENTS

I. INTRODUCTION

This chapter describes the utilization of an animal model to identify hair growth drug candidates. The limited access to human or monkey models suitable for the study of hair growth makes the development and proper utilization of other *in vivo* systems important. While several differences exist between human and rodent models, they have in common a dermal hair follicle matrix and cycling hair follicles that can be influenced by internal or external variables.[1] In human scalp, there exists at all times a mixture of actively growing (anagen) or resting (telogen) follicles, while most rodents transition through well-characterized age-dependent, synchronized follicle cycles. The mouse displays hair cycle characteristics that are well suited for laboratory experimentation aimed toward development of drugs with trichogenic potential.

Several aspects of hair growth can be investigated in animal models with synchronized follicle cycles. Among these are the premature induction of anagen during the telogen cycle and the prolonged maintenance of the anagen cycle.[2] In addition, formulation variables can be tested for their influence on drug delivery and efficacy. The mouse model presented here is suitable for the screening of potential hair growth compounds prior to human clinical testing.

II. DESCRIPTION OF THE TEST SYSTEM

The C3H mouse has brown hair and a dorsal skin that is thin and pink in color during the telogen cycle, and that darkens with subcutaneous follicle bulb pigmentation during early anagen. This darkening of the skin allows for easy determination of anagen onset. The hair follicles of these animals cycle between anagen, catagen (the follicle involution phase), and telogen at regular intervals dependent primarily on age, however, this cycling is somewhat influenced by seasonal or environmental changes as well. Figure 1 shows the basic pattern of anagen and telogen hair growth cycles in the mouse.

The telogen 2 cycle provides an extended period of time (approximately 50 days) during which drugs can be screened for their ability to promote a premature third anagen cycle, also referred to as a shortening of the length of the telogen 2 cycle. Atraumatic removal of hair from a mid-dorsal test site allows for evaluation of hair growth effects using dermal injections or topical application of test articles.

III. PEPTIDE–COPPER COMPLEXES
FOR HAIR GROWTH

Glycyl-L-histidyl-L-lysine (GHK, single letter amino acid abbreviations) was first described as a tripeptide isolated from human serum that enhanced the growth and viability of a variety of cultured cells.[3-6] The tripeptide was isolated from Cohn Fraction V albumin fraction of human serum and has been shown to

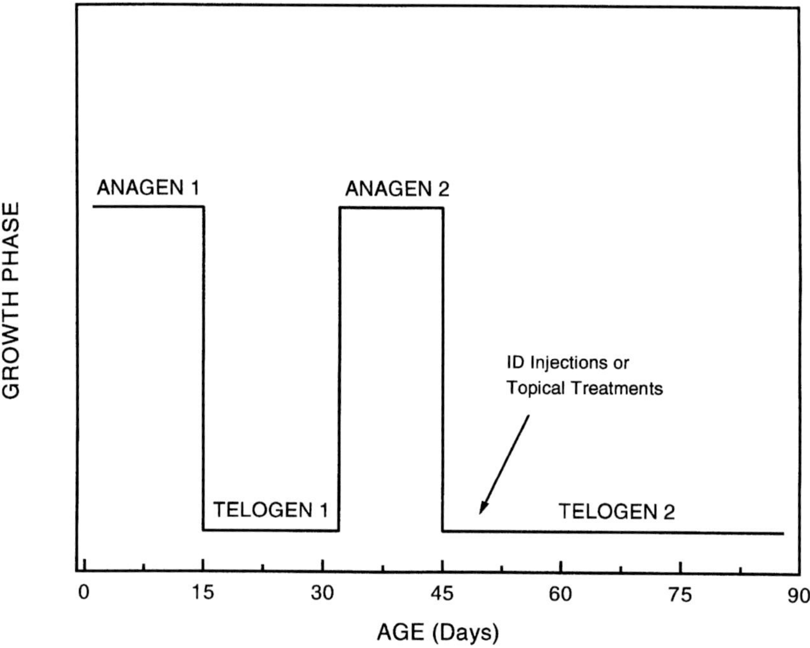

FIGURE 1 Hair follicle cycles in mice. The hair follicle cycles of mice follow a synchronized, primarily age-dependent pattern as shown in this figure. While the follicle involution phase (catagen) is not shown, it occurs in the transition between the anagen (growing) and telogen (resting) cycles. In the studies presented here, administration of drug was made during the long telogen 2 cycle (approximately age 50 to 100 days).

coisolate as a complex with copper, designated as GHK–Cu.[7,8] Subsequent studies have shown that the GHK–Cu complex possesses biological activities that indicate it may be an important modulator of the tissue repair process. For example, GHK has been shown to be a potent chemoattractant for cells critical to the tissue restoration phase of the wound-healing process. GHK incorporated into an agarose gel assay was found to be a potent chemoattractant for isolated rat peritoneal mast cells[9] and GHK incorporated in a polymer (Elvax)/dacron device implanted in the backs of groups of mice was shown to be a chemoattractant for macrophages, monocytes, and mast cells.[10] In studies with subcutaneously implanted stainless-steel chambers in rats, histopathological analysis early in the wound-healing process showed more inflammatory cells (polymorphonuclear cells and macrophages) in the GHK–Cu-treated chambers compared to the saline controls.[11] Angiogenesis or new blood vessel growth is an important aspect of the tissue repair and maintenance process. Copper complexed to GHK has been shown to be angiogenic in both *in vitro* and *in vivo* models.[12,13]

The stimulation of the synthesis of various components of the extracellular matrix is also a major biochemical property of peptide–copper complexes. The extracellular matrix has been implicated in the regulation of hair growth.[14] Maquart et al.[11,15,16] reported that GHK–Cu treatment stimulated the synthesis of both collagen and glycosaminoglycans (GAGs) both *in vitro* and *in vivo*. Many analogs of the GHK–Cu structure, including those shown to be active in hair growth models described here and in the accompanying manuscript, are active in stimulating the synthesis of extracellular matrix components (unpublished observations).

The hair growth drug candidates used in the studies presented here were developed from observations made during the course of rodent wound-healing studies in which an analog of the biologically active GHK–Cu complex was used. Peptide analog development for hair growth has generally followed the path to the synthesis of compounds with enhanced activity in the intradermal injection hair growth model in mice presented below. All of the analogs in this series have been composed of specific short peptides, 3 to 7 naturally occurring amino acid residues, at a ratio of 1 to 2 mol of peptide to 1 mol of copper ion.

Percutaneous absorption studies comparing a number of peptide–copper complex hair growth drug candidates have shown that there is a strong relationship of peptide structure and size to the ability of the compounds to penetrate the skin barrier layers. In addition, activity assays have also indicated correlations between structure and the types of topical formulations that are needed to elicit a positive hair growth response. Successful drug candidates for hair growth must combine both intrinsic hair growth-promoting activity (detected in the intradermal injection model) and the ability to elicit hair growth by topical application.

To evaluate the effectiveness of potential hair growth drug candidates, example are shown here in which several potential candidates and control compounds were tested in the following experimental models. These included peptide–copper complexes, a peptide without copper, and copper alone.

IV. EXPERIMENTAL METHODS

A. ANIMALS

Female C3H mice (Simonsen Labs, Gilroy, CA) were used in these studies. Animals were received at approximately 50 days of age and housed in cages three to five per cage. Food and water were provided *ad libitum* throughout the study. All animals were entered into these studies after confirmation of the telogen cycle by carefully removing the dorsal hair using clippers and observation of the pink skin.

B. INTRADERMAL INJECTION

Female C3H mice ($n = 110$; telogen 2 cycle, age 50 to 60 days) were randomly assigned to treatment groups ($n = 5$ to 8 per group). Two dorsal sites within the clipped region were identified (cranial and caudal) and intradermal injections of the test article (0.1 ml) were preformed using a sterile 30-gauge needle.

C. TOPICAL ADMINISTRATION

Female C3H mice ($n = 110$; telogen 2 cycle, age 50 to 60 days) were randomly assigned to treatment groups ($n = 5$ to 8 per group). Dorsal hair was carefully removed using clippers and confirmation of telogen was made (pink skin). Prior to study initiation, test articles (except minoxidil) were formulated in a vehicle consisting of propylene glycol, ethanol, water, and Nonoxynol-9. Treatments (of approximately 0.1 ml) were performed twice per day (Monday to Friday) to the clipped dorsal region (approximately 4 cm^2, caudally based) using cotton-tipped applicators. Treatments began during the telogen 2 cycle (approximately age 50 to 55 days). Topical treatments were continued until gross evidence of follicle pigmentation was observed, indicating the start of the anagen phase.

D. QUANTIFICATION OF RESPONSE

1. Response Latency
    (Topical Studies Only)

Response latency (anagen onset) was determined by recording the follicular cycle (anagen or telogen) on a daily basis, and counting the number of days until evidence of follicle pigmentation was seen. Pigmentation is a well-defined and grossly visible marker for anagen. Group means and SEM (standard error of the mean) were calculated using individual animal response latencies.

2. Degree of Irritation

A rating scale was used for both injection and topical studies. The level of irritation represents the maximum observed irritation independent of time.

Injection Study

0 = none
1 = slight (reddening)
2 = moderate (blanching or small open lesion <1 mm^2)
3 = severe (open lesion area >1 mm^2)

Topical Study

0 = none
1 = slight (drying, flaking skin)
2 = moderate (reddening of skin with peeling)
3 = severe (epidermal bleeding, blistering, scabbing)

3. Growth Area Determination

Hair growth area was measured using computer-based image analysis (Bioquant™). For the injection studies (days 15 and 21), the two independent

TABLE 1
Test Articles

Test article code	Structure	Peptide:copper molar ratio	Molecular weight	Test system
PC1234	AHK-Cu[a]	1.1:1	418	Inj. and top.
PC1031	GHKVFV-Cu[b]	2:1	1435	Inj. and top.
PC1024	GGHK-Cu[c]	2:1	859	Inj. only
PC1048	AHK[d]	—	354	Inj. only
PC1154	Copper acetate	—	200	Inj. only
Minoxidil	—	—	200	Top. only

Note: Test articles used in intradermal injection (Inj.) and topical (top.) administration studies. In the injection study, three peptide–copper complexes (PC1234, PC1031, and PC1024) were included for comparison with a peptide without copper (PC1048) and copper acetate (PC1154). In the topical study, PC1234 and PC1031 were used for comparison with minoxidil (Rogaine™, 2%).

[a] L-Alanyl-L-histidyl-L-lysine copper complex.

[b] Glycyl-L-histidyl-L-lysyl-L-valyl-L-phenylalanyl-L-valine copper complex.

[c] Glycyl-glycyl-L-histidyl-L-lysine copper complex.

[d] L-Alanyl-L-histidyl-L-lysine.

injections were measured, or if grown together, the total area was measured and divided by two. Individual animal data were used to calculate the mean response area. For topical studies (days 15, 21, and 28), the entire treatment area (clipped area) as well as the area of induced hair growth was measured and these were used to calculate the percentage treatment area response ([area of growth/total area] × 100).

E.   TEST ARTICLES

Table 1 describes the test articles used in these studies. These compounds consist of peptide–copper complexes of varying structure formulated at a molar ratio of either 1:1 or 2:1 peptide to copper. Two additional test articles were used to determine the contribution of individual components to the activity, the AHK peptide without copper and copper acetate.

V.   TEST RESULTS AND PRESENTATION OF DATA

A.   INJECTION STUDY

Results from the intradermal injection study are provided in Table 2. The first indication of a positive hair growth response was typically seen between

TABLE 2
Injection Study Results

Test article	Dose (μmol per injection)	Response rate (*n* positive/total *n*)	Degree of irritation	Response area — cm² (mean ± SEM) Day 15	Day 21
PC1234	2.39	6/7[a]	1	3.3 ± 0.4	3.9 ± 0.5
	1.20	7/8	0	1.9 ± 0.3	2.4 ± 0.3
	0.24	2/8	0	0.1 ± 0.0	0.3 ± 0.2
PC1031	0.67	8/8	0	1.6 ± 0.2	2.0 ± 0.2
	0.34	8/8	0	0.5 ± 0.1	1.1 ± 0.2
	0.07	1/8	0	0.0 ± 0.0	0.0 ± 0.0
PC1024	2.33	1/8	0	0.3 ± 0.2	0.4 ± 0.3
	1.16	1/8	0	0.0 ± 0.0	0.0 ± 0.0
	0.58	0/8	0	0.0 ± 0.0	0.0 ± 0.0
PC1048	2.82	0/5	0	0.0 ± 0.0	0.0 ± 0.0
	1.41	0/5	0	0.0 ± 0.0	0.0 ± 0.0
	0.28	0/5	0	0.0 ± 0.0	0.0 ± 0.0
PC1154[b]	2.21	5/5	3	1.5 ± 0.3	2.3 ± 0.4
	1.10	5/5	3	1.6 ± 0.3	2.1 ± 0.3
	0.22	5/5	2	0.2 ± 0.0	0.4 ± 0.0
Saline	0.00	0/8	0	0.0 ± 0.0	0.0 ± 0.0

Note: Results from the injection study include comparisons of response rate (animals with a positive hair growth response), degree of irritation (based on rating scale provided in the methods section), and response area as measured by image analysis. Date presented are mean ± SEM. PC1154 (copper acetate), caused extensive irritation and skin ulceration. The site of ulceration was included in the measurement area (see text).

[a] One animal expired unrelated to treatment.

[b] Response area includes ulcerated areas.

10 and 14 days after injection of the test article. In this study, the peptide–copper complexes PC1234 (a tripeptide copper complex) and PC1031 (a hexapeptide copper complex) were effective promoters of hair growth. Dose-dependent responses were seen in the response rate and in the area of hair growth. Equimolar dose comparisons were not made, however, nearly equivalent growth activity was seen when PC1031 was administered at approximately one-half the dose of PC1234 (0.67 vs. 1.20 μmol/injection). Both PC1234 and PC1031 caused minimal to no observable irritation at the injection sites.

A representative photograph of a mouse injected at the two sites with PC1031 and the corresponding photomicrograph showing the large anagen phase hair follicles at the injection site is shown in Figure 2. Hair growth is present only at the sites of injection while the surrounding clipped area remains in telogen. A similar result is observed in the photomicrograph. Large anagen

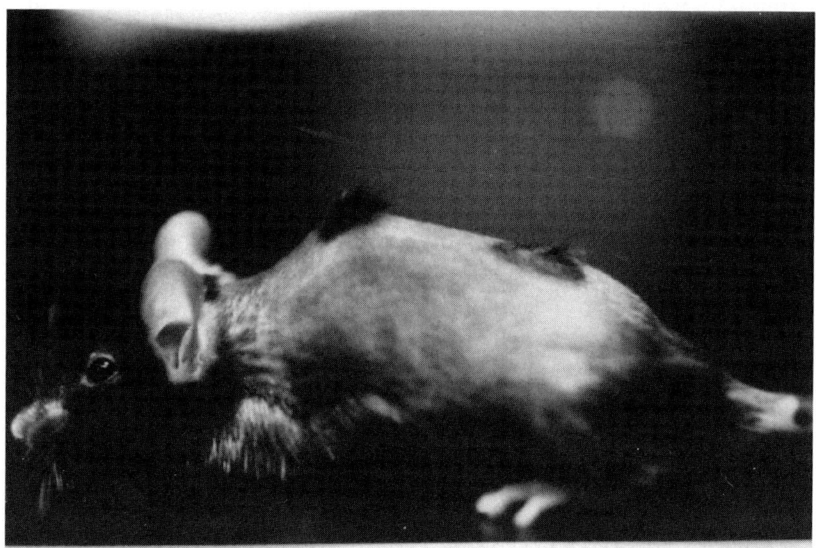

A

B

FIGURE 2 Typical hair growth following intradermal injection of peptide–copper complexes. A representative photograph of a mouse injected at the two sites with PC1031 (A) and the corresponding photomicrograph showing the large anagen phase hair follicles at the injection site (B). Hair growth is present only at the sites of injection while the surrounding clipped area remains in telogen. Large anagen hair follicles are present at the site of injection (centered at the right margin of the photomicrograph) while the follicles distant from the injection site (middle to left margin of photomicrograph) remain in telogen.

hair follicles are present at the site of injection (centered at the right margin of the photomicrograph) while the follicles distant from the injection site (middle to left margin of the photomicrograph) remain in telogen.

PC1154 (copper acetate) was tested to determine if the hair growth activity was due only to the copper moiety of the complexes. Intradermal injection of PC1154 (Table 2) caused extensive tissue damage at the doses tested (comparable to the copper content of the PC1234 doses). Data presented for this compound are also biased due to inclusion of the ulcerated injection site in the measured hair growth area.

The contribution of the peptide moiety to the hair growth activity was tested with two additional compounds. The tetrapeptide–copper complex, PC1024, a peptide–copper complex of similar structure to the active complexes, and the tripeptide compound, PC1048 (the peptide moiety of the PC1234 complex, AHK), and the saline vehicle were all found to be inactive at the doses tested.

Comparison of relative hair growth activity of these compounds on study day 21 is shown in Figure 3. PC1031 was the most potent compound in this model via injection. The PC1154 produced a large amount of irritation and ulceration at the injection site, which was included in the apparent area of hair growth. Both PC1031 and PC1234 were more potent than PC1154 at the highest dose, showing a clear difference in activity of the peptide–copper complex when compared to copper alone.

B.   TOPICAL STUDY

The active compounds from the intradermal study, PC1234 and PC1031, were included in a topical administration study for comparison with minoxidil (Rogaine™) and appropriate control groups (vehicle control consisted of the same topical formulation as used with PC1234 and PC1031). As described previously, evaluation of efficacy was based on response rate, response latency, and degree of response (area). In addition, a determination of irritation was made using the scale as described for topical studies.

As shown in Table 3, PC1234, PC1031, and minoxidil were effective in promoting hair growth via topical administration (response rate). Animals treated with these compounds showed a response (latency) within the 28-day study period. Both control groups were unresponsive throughout the course of the study. PC1234 was most efficacious with respect to response latency (day 9 for 1.25 and 0.67%), compared to approximately 21 days for PC1031 and 11 days for minoxidil. Both peptide–copper complexes were more irritating than either minoxidil or the vehicle alone. Moderation of the irritation was observed at the lower doses of both compounds.

Data obtained by analysis of the hair growth response are presented in Table 4. Measurements were made on study days 15, 21, and 28 (end of study). PC1234 treatment (1.25 and 0.67%) was the most potent compound as determined by measurement of hair growth within the treatment area. By study

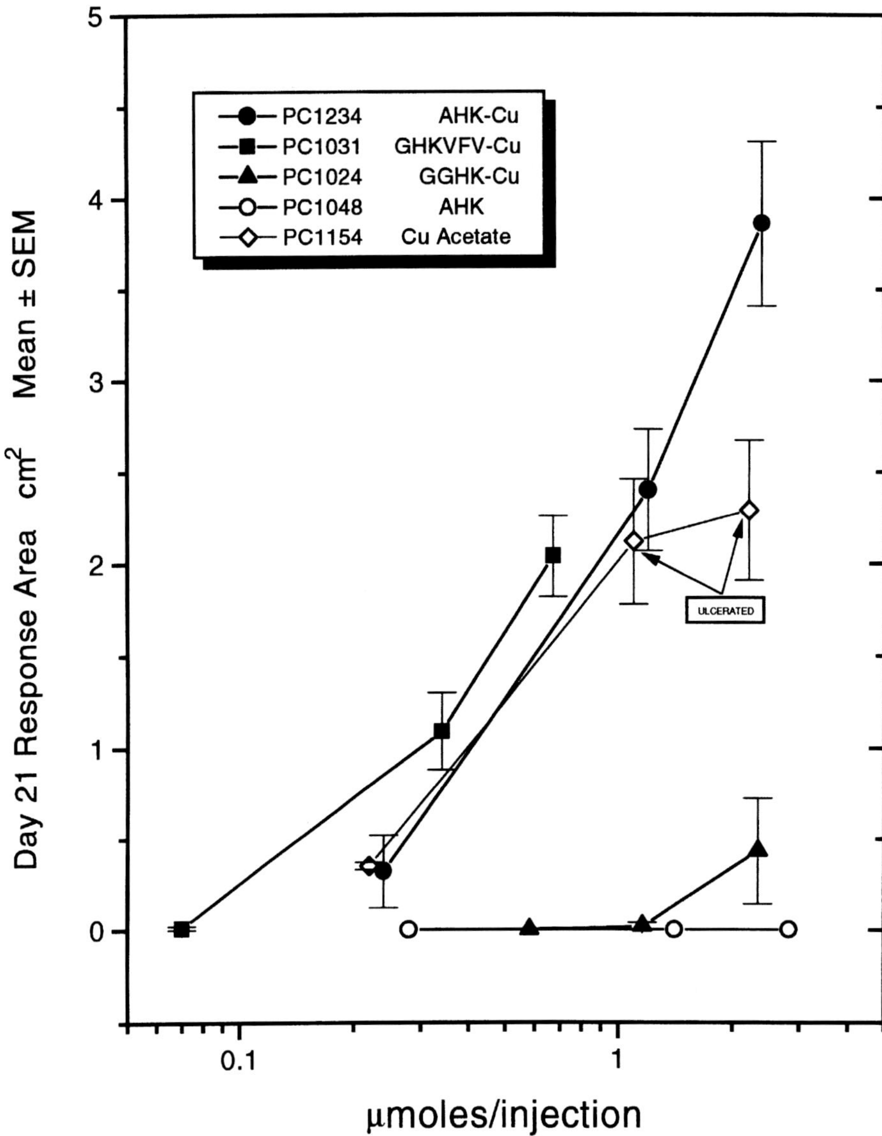

FIGURE 3 Hair growth following intradermal injection of peptide–copper complexes. Day 21 areas of hair growth following intradermal injection of the indicated test compounds. The data shown represent the mean ± SEM of the cm^2 of hair growth.

day 15, 100% of the treatment area was in anagen, compared to 0% for comparable PC1031 doses, and 33% for minoxidil at this time point. On day 21, shown in Figure 4 for comparison with the injection results, a treatment area response of approximately 35% was measured in the low-dose PC1234 group (0.25%). PC1031 treatment (5 and 2.5%) caused a 53 to 56% response, while minoxidil treatment induced a 51% response. The PC1234 response at

TABLE 3
Rate, Latency, and Irritation

Test article	Dose conc. (%)	Response rate (*n* positive/total *n*)	Degree of irritation	Response latency days (mean ± SEM)
PC1234	1.25	8/8	2	9.0 ± 0.0
	0.67	8/8	2	9.0 ± 0.0
	0.25	8/8	1	18.4 ± 1.2
PC1031	5.00	8/8	1.5	21.0 ± 0.0
	2.50	8/8	1.5	17.9 ± 0.9
	1.00	7/7[a]	1	20.5 ± 1.4
Minoxidil	2.00	7/8	0	11.1 ± 1.8
Vehicle	0.00	0/8	0	No growth
Untreated	—	0/8	0	No growth

Note: Results from the topical study with respect to response rate (animals with a positive hair growth response), response latency (number of treatment days to follicle pigmentation), and degree of irritation (based on rating scale provided in the methods section). Data represent the mean ± SEM.

[a] One animal expired unrelated to treatment.

TABLE 4
Growth Response

Test article	Dose conc. (%)	Growth response — percent of treated area (mean ± SEM)		
		Day 15	Day 21	Day 28
PC1234	1.25	100 ± 0	100 ± 0	100 ± 0
	0.67	100 ± 0	100 ± 0	100 ± 0
	0.25	0 ± 0	34.6 ± 15.0	94.0 ± 4.0
PC1031	5.00	0 ± 0	53.3 ± 15.5	93.5 ± 4.2
	2.50	0 ± 0	56.6 ± 14.8	98.0 ± 1.1
	1.00	0 ± 0	14.2 ± 9.6	68.8 ± 10.3
Minoxidil	2.00	33.4 ± 12.0	51.9 ± 13.3	63.8 ± 12.7
Vehicle	0.00	0 ± 0	0 ± 0	0 ± 0
Untreated	—	0 ± 0	0 ± 0	0 ± 0

Note: Growth response from topical drug treatment is expressed as percentage of the treated area [(area of growth/total area) × 100]. Data represent the mean ± SEM obtained using image analysis on days 15, 21, and 28.

the 0.25% dose level was 94% on day 28, compared to 64% for minoxidil, and 93 to 98% for the mid and high dose PC1031 groups. Both control groups remained in telogen for the duration of the study.

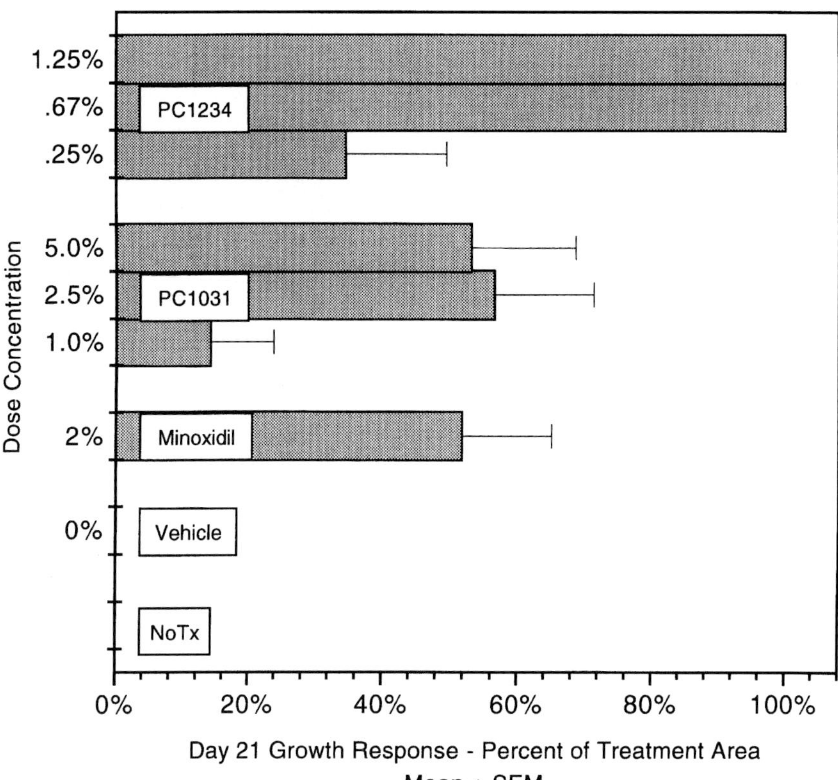

FIGURE 4   Hair growth following topical application of peptide–copper complexes. Day 21 of treated area showing hair growth following twice daily (five times/week) application of the indicated test compounds. The data shown are the mean ± SEM of the percentage of the total treated area showing hair growth on day 21 of the study.

VI. INTERPRETATION OF THE RESULTS

An *in vivo* hair growth model has been described that aids in the identification of potential effective hair growth drugs. Quantitative analyses of efficacy have been provided with respect to intradermal injection and topical administration of peptide–copper complexes. An essential comparison to minoxidil allows for interpretation of the relative activity of these compounds using topical administration. The important conclusions from these studies were as follows:

1. PC1234 and PC1031 are potent stimulators of the transition of telogen follicles to anagen follicles when used intradermally or topically. Minoxidil treatment was also effective when used topically. A predominant feature of androgenetic alopecia is the presence in increasing numbers of telogen follicles. Based on that characteristic, the effect of these compounds on telogen follicles is a significant finding.

2. Differences in the potency of peptide–copper complexes could be observed in the intradermal injection model. While both PC1234 and PC1031 showed efficacy in this model, another peptide–copper complex of similar structure (PC1024) was ineffective.
3. Both copper acetate (PC1154) and a peptide without copper (PC1048) produced unacceptable results in the injection model. While PC1154 caused extensive local irritation and ulceration, PC1048 showed no hair growth activity. These results, combined the inactivity of PC1024, show that the activity of peptide–copper complexes is dependent upon both the copper component and the proper peptide sequence. The peptide may serve to deliver the copper to the site of action in a less toxic manner and allow use of pharmacologically active doses without the irritation associated with copper salts.
4. The hexapeptide–copper complex PC1031, while more potent when administered intradermally, is less potent when used topically than the tripeptide–copper complex PC1234.

Topical administration studies serve as a critical evaluation tool when developing hair growth drugs from the pharmacologic, toxicologic, and kinetic perspectives. The desired clinical method of delivery would be topical rather than injection. However, these studies tend to be more labor intensive, take a longer period of time to perform, and require more compound, and the results are more difficult to interpret with respect to determining actual dose delivery through the skin. Intradermal injection studies allow for effective elimination of less potent compounds, require little test material, provide for a controlled dose presentation, and generate meaningful data in a relatively short period of time. When used in concert, both test systems serve to adequately characterize the effect of a drug candidate on telogen hair follicles.

References

1. **Ebling, F.J.,** Hair, *J. Invest. Dermatol.,* 67, 98, 1976.
2. **Uno, H., Schroeder, B., Fors, T.D., and Mori, O.,** Macaque and rodent models for the screening of drugs for stimulating hair growth, *Ann. N.Y. Acad. Sci.,* 642, 470, 1991.
3. **Pickart, L. and Thaler, M.,** Tripeptide in human serum which prolongs survival of normal liver cells and stimulates the growth of hepatoma cells, *Nature New Biol.,* 243, 85, 1973.
4. **Schlesinger, D.H., Pickart, L., and Thaler, M.M.,** Growth modulating tripeptide is H-Gly-His-Lys-OH, *Experientia,* 33, 324, 1977.
5. **Pickart L., Thayer, L., and Thaler, M.,** A synthetic tripeptide which increases survival of normal liver cells and stimulates the growth of hepatoma cells, *Biochem. Biophys. Res. Commun.,* 54, 562, 1973.

6. **Pickart, L.,** The use of glyclyhistidyllysine in culture systems, *In Vitro*, 17, 459, 1981.
7. **Pickart, L., Thaler, M.M., and Millard, M.M.,** Effect of transition metals on recovery from plasma of the growth modulating tripeptide glyclyhistidyllysine (GHL), *J. Chromatogr.*, 175, 65, 1979.
8. **Pickart, L. and Thaler, M.M.,** Growth-modulating tripeptide (glycylh-istidyllysine): association with copper and iron in plasma, and stimulation of dhesiveness and growth of hepatoma cells in culture by tripeptide-metal ion complexes, *J. Cell. Physiol.*, 102, 129, 1980.
9. **Poole T.J. and Zetter, B.R.,** Stimulation of rat peritoneal mast cell migration by tumor derived peptides, *Cancer Res.*, 43(12), 5857, 1983.
10. **Zetter, B.R., Rasmussen, N., and Brown, L.,** An in vivo assay for chemoattractant activity, *Lab. Invest.*, 53(3), 362, 1985.
11. **Maquart, F.-X., Bellon, G., Chaqour, B., Wegrowski, J., Patt, L.M., Trachy, R.E., Monboisse, J.-C., Chastang, F., Birembaut, P., Gillery, P., and Borel, J.-P.,** In vivo stimulation of connective tissue accumulation by the tripeptide copper complex glycyl-L-histidyl-L-lysine-Cu2+ in rat experimental wounds, *J. Clin. Invest.*, 92, 2368, 1993.
12. **Raju K.S., Alessandri, G., and Gullino, P.M.,** Characterization of a chemoattractant for endothelium induced by angiogenesis factors, *Cancer Res.*, 44(4), 1579, 1984.
13. **Raju, K.S., Alessandri, G., Ziche, M., and Gullino, P.M.,** Ceruloplasmin, copper ions, and angiogenesis, *JNCI*, 69(5), 1183, 1982.
14. **Messenger, A.G.,** Extracellular matrix and the hair growth cycle, *J. Invest. Dermatol.*, 96, 75s, 1991.
15. **Maquart, F.X., Pickart, L., Laurent, M., Gillery, P., Monboisse, J.C., and Borel, J.P.,** Stimulation of collagen synthesis in fibroblast cultures by the tripeptide-copper complex glycyl-L-histidyl-L-lysine-Cu++, *FEBS Lett.*, 238(2), 343, 1988.
16. **Wegrowski, Y., Maquart, F.X., and Borel, J.P.,** Stimulation of sulfated glycosaminoglycan synthesis by the tripeptide-copper complex glycyl-L-histidyl-L-lysine-Cu 2+, *Life Sci.*, 51(13), 1049, 1992.

19 Powdered Human Stratum Corneum: Model for Interaction of Chemical and Human Skin

Ronald C. Wester, Xiaoying Hui, and Howard I. Maibach

TABLE OF CONTENTS

I. INTRODUCTION

The human stratum corneum includes the horny pads of the palms and soles (callus) and the membranous stratum corneum covering the remainder of the body.[1] It represents a rate-limiting barrier in the transport of most chemicals across the skin.[2-5] Chemicals must first partition into the stratum corneum before entering deeper layers of the stratum corneum, epidermis, and dermis to reach the vascular system. This partitioning process occurs much faster than complete diffusion through the whole stratum corneum, and the process quickly reaches equilibrium.[6] In addition to binding within the stratum corneum, a chemical can also be retained within the stratum corneum as a reservoir.[7] Thus, understanding the process of chemical partitioning into the stratum corneum becomes important in developing an insight into the barrier properties and transport mechanism.

Powdered human stratum corneum (PHSC) is a product made from foot callus that a podiatrist will routinely remove and discard. The foot callus is cut into smaller pieces and ground with dry ice to form a powder. Uniformity in particle size is achieved with sieving. In our laboratory, that portion of the powder that passed through a 40-mesh but not an 80-mesh sieve was used. The callus is human stratum corneum derived and, thus, should retain some physical and chemical characteristics of human stratum corneum.[8]

II. PHSC PARTITIONING AND PERCUTANEOUS ABSORPTION

Table 1 shows the partition of benzene, 54% PCBs, and nitroaniline between water and PHSC. The more lipophilic PCBs were the higher binding compound, followed by benzene and then the more hydrophilic nitroaniline. Thus, the binding/partitioning is able to distinguish between compounds. The ten fold binding of nitroaniline concentrations also shows linearity. Table 2 shows that the binding of nitroaniline to PHSC correlated well with nitroaniline *in vitro* percutaneous absorption in human skin and *in vivo* percutaneous absorption in the Rhesus monkey.[8]

III. PHSC AND SKIN DECONTAMINATION

PHSC can be used to determine which chemicals/decontaminants might be able to remove (decontaminate) hazardous chemicals from human skin. The procedure is to mix a chemical into the PHSC and then try to remove that

---

TABLE 1

**Partition of Contaminants from Water Solution
to Powdered Human Stratum Corneum
following 30-min Exposure**

	Chemical Contaminant (µg/ml)	Percent dose partitioned to skin
Benzene	21.7	16.6 ± 1.4
54% PCB	1.6	95.7 ± 0.6
Nitroaniline	4.9	2.5 ± 1.1
	1.6	3.7 ± 0.7
	0.49	3.9 ± 1.5

TABLE 2

***In Vivo* Percutaneous Absorption of *p*-Nitroaniline in the Rhesus Monkey following 30-min Exposure to Surface Water: Comparisons to *In Vitro* Binding and Absorption**

Phenomenon	Percent dose absorbed/bound
In vivo percutaneous absorption, Rhesus monkey	4.1 ± 2.3
In vivo percutaneous absorption human skin	5.2 ± 1.6
In vitro binding, powdered human stratum corneum	2.5 ± 1.1

chemical with a series of possible decontaminants. The liquid decontaminant is poured into the chemical/PHSC and mixed. After a set time period centrifugation will separate a solution from the PHSC. The content of the solution will show the decontamination potential of the solution. This is shown in Table 3 and Figure 1, where alachlor readily binds to PHSC. Water alone will remove only a small portion of the alachlor. However, the introduction of 10% soap removes a larger portion of the alachlor and 50% soap removes most of the alachlor. Perhaps this is an elegant way to show that soap in water is necessary to wash one's hands. However, it does illustrate the adaptability of the PHSC to demonstrate skin decontamination.[3]

IV. PHSC AND DISEASED SKIN

Powdered human stratum corneum does have the potential for medical research application. For instance, a set of vehicles can be screened to determine which vehicle most readily releases a drug into the stratum corneum. This

TABLE 3

Partitioning: Alachlor in Lasso with Powdered Human Stratum Corneum

	[^{14}C]Alachlor (% dose)
Stratum corneum	90.3 ± 1.2
Lasso supernatant	5.1 ± 1.2
Water only wash of stratum corneum	4.6 ± 1.3
10% soap and water wash	77.2 ± 5.7
50% soap and water wash	90.0 ± 0.5

Note: [^{14}C]Alachlor in Lasso EC formulation (1:20 dilution) mixed with powdered human stratum corneum, let set for 30 min, then centrifuged. Stratum corneum wash with (1) water only, (2) 10% soap and water, and (3) 50% soap and water.

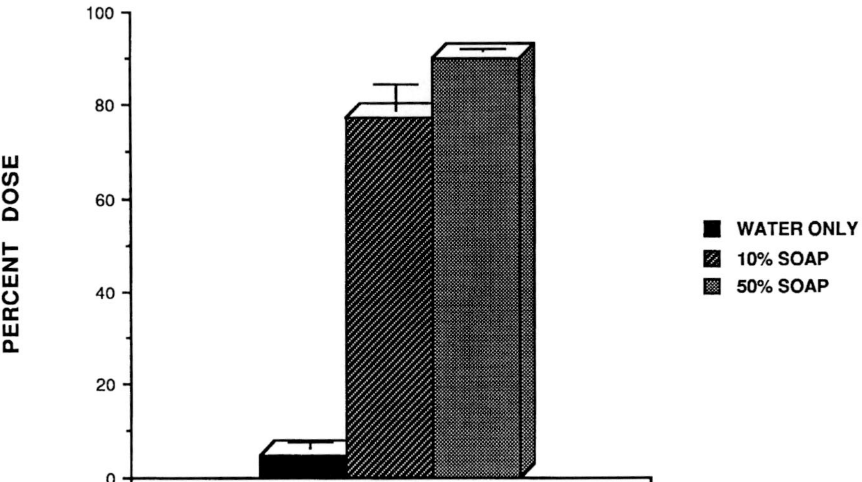

FIGURE 1   Alachlor will bind to powdered human stratum corneum (as it will bind to human skin). PHSC can compare decontamination methods. Here, soap is needed to remove alachlor from skin (plain water will not work).

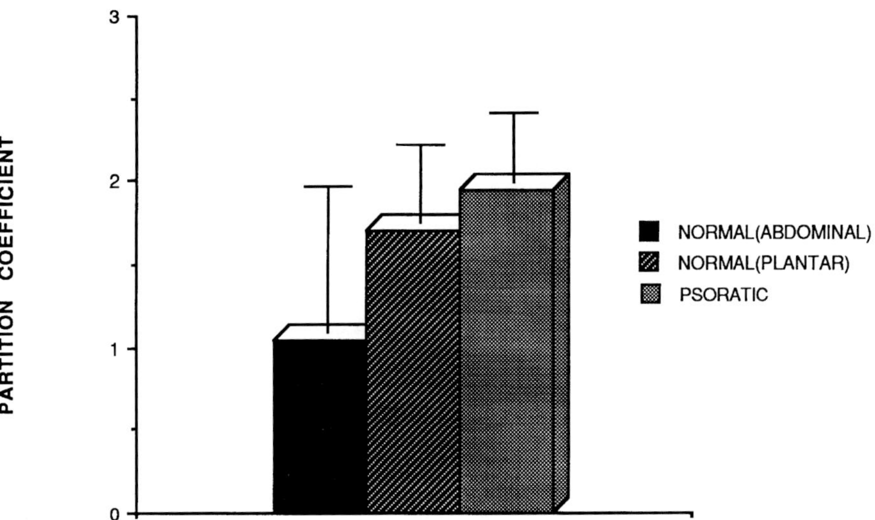

FIGURE 2   Drugs need to be in contact with diseased skin at the site of action. Here, the study showed that hydrocortisone will bind with psoriatic PHSC (as well as normal PHSC).

would help with drug delivery into the skin. Also, diseases involving the stratum corneum can be studied using PHSC. An example in Figure 2 is the partitioning of hydrocortisone from normal and psoriatic PHSC. In this case there was no difference in partitioning between normal and psoriatic PHSC.[9] It should be noted that there is no difference in the *in vivo* percutaneous absorption of hydrocortisone in normal volunteers and psoriatic patients.[10]

TABLE 4

Partition Coefficients of Arsenic[a] and Cadmium Chloride[b] in Powdered Human Stratum Corneum/Water and Soil/Water

Test material	Partition coefficient
Arsenic	
Stratum corneum	1.1×10^4
Soil	2.4×10^4
Cadmium Chloride	
Stratum corneum	3.61×10^1
Soil	1.03×10^5

[a] $\text{Partition coefficient} = \dfrac{\text{Concentration of arsenic} - 73 \text{ in } 1000 \text{ mg PHSC (soil)}}{\text{Concentration of arsenic} - 73 \text{ in } 1000 \text{ ml water}}.$

[b] Partition coefficient = Concentration of $CdCl_2$ in 1000 mg of powdered human stratum corneum (soil)/Concentration of $CdCl_2$ in 1000 mg of water.

V. PHSC AND ENVIRONMENTAL HAZARDOUS CHEMICALS

The PHSC will readily separate from any liquid with centrifugation. Thus, it is readily adaptable to a variety of liquid solutions. One of our laboratory's interests is the potential percutaneous absorption of contaminants from soil. Soil can be readily mixed with PHSC, but centrifugation will not separate the two. Something that can be done is to determine separate partition coefficients relative to a common third liquid. This is shown for cadmium and arsenic in Table 4. These "relative" partitions can then be compared to those of other compounds and to skin absorption values.[11,12]

VI. PHSC AND QSAR PREDICTIVE MODELING

Many experiments have been conducted to predict chemical partitioning into the stratum corneum *in vitro*. However, most were based on quantitative structure–activity relationships (QSARs) or related chemicals to determine the partitioning process, and few studies focus on structurally unrelated chemicals.[13] Since the range of molecular structure and physicochemical properties is very broad, any predictive model must address a broad scope of partitioning behavior.

This study assesses the relationship of a number of chemicals with a broad scope of physicochemical properties in the partitioning mechanism between the PHSC and water. Uniqueness and experimental accuracy are added by using the PHSC. The experimental approach is designed to determine how the

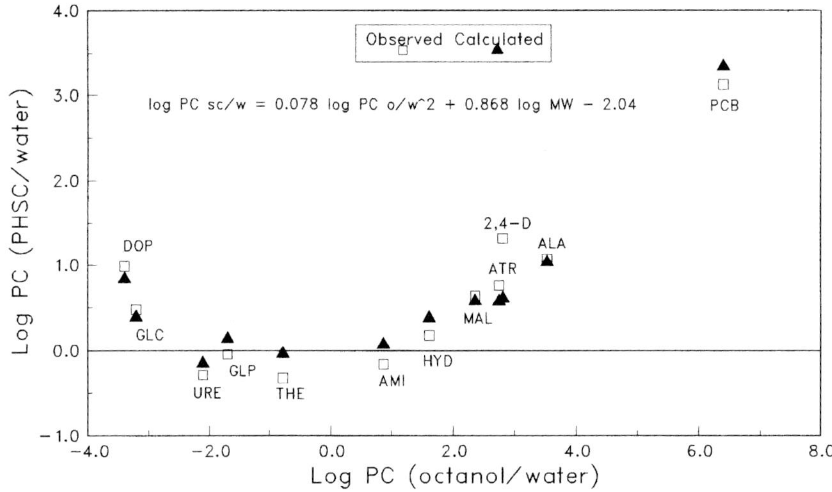

FIGURE 3 Correlation of the log PC (PHSC/w) and the log PC (o/w) of the 12 test chemicals. Data of PC (PHSC/w) of each chemical were the equilibrium or the maximum value with 24 h of incubation cited from Figures 2 or 3. Open symbols express observed values and each represents the mean of a test chemical ±SD (n = 5). Closed symbols express values calculated. DOP, dopamine; GLC, glycine; URE, urea; GLP, glyphosate; THE, theophylline; AMI, aminopyrine; HYD, hydrocortisone; MAL, malathion; ATR, atrazine; 2,4-D, 2,4-dichlorophenoxyacetic acid; ALA, alachlor; PCB, polychlorinated biphenyls.

PC PHSC/w is affected by (1) chemical concentration; (2) incubation time; (3) chemical lipophilicity (or hydrophilicity), and other factors. These parameters are used to develop an *in vitro* model that will aid in predicting chemical dermal exposure for hazardous chemicals.

Figure 3 describes a smooth, partially curvilinear relationship between the log PC (PHSC/S) and the log PC (o/w) of many chemicals. Lipophilicities and hydrophilicities of compounds were defined as log PC (o/w) larger or smaller than 0, respectively. For lipophilic chemicals, such as aminopyrine, hydrocortisone, malathion, atrazine, 2,4-D, alachlor, and PCB the logarithms of PHSC/water partition coefficients are proportional to the logarithms of the octanol/water (o/w) partition coefficients.

$$\log PC\,(PHSC/w) = 0.59\,\log PC\,(o/w) - 0.72$$

Student's t value :   9.93

$$n = 7, \quad r^2 = 0.95, \quad S = 0.26, \quad F = 98.61$$

For hydrophilic chemicals, such as theophylline, glyphosate, urea, glycine, and dopamine, the log PC (PHSC/W) values are approximately and reversely proportional to log PC (o/w).

$$\log PC\,(PHSC/w) = -0.60\ \log PC\,(o/w) - 0.27$$

$$\text{Student's } t \text{ value}: \quad -4.86$$

$$n = 5, \quad r^2 = 0.88, \quad S = 0.26, \quad F = 23.61$$

However, the overall relationship of the PC (PHSC/w) of these chemicals and their PC (o/w) is nonlinear. This nonlinear relationship is adequately described by the following equation:

$$\log PC\,(PHSC/w) = 0.078\ \log PC\,(o/w)^2 + 0.868\ \log MW - 2.04 (5)$$

$$\text{Student's } t \text{ value}: \quad 8.29,\ 2.40$$

$$n = 12, \quad r^2 = 0.90, \quad S = 0.33, \quad F = 42.59$$

The logarithm of molecular weight (MW) gave a stronger correlation in this regression than MW ($t = 1.55$). In Figure 3, the calculated log PC (PHSC/w) (Y estimate) are compared to the corresponding observed values for these chemicals. As shown, the calculated values are acceptably close to the observed values. The correspondence with minimal scatter suggests that this equation would be useful in predicting *in vitro* partitioning in the PHSC for important environmental chemicals.

VII. DISCUSSION

Human stratum corneum has been used as an *in vitro* model to explore percutaneous absorption and risk of dermal exposure.[14,15] The traditional method of preparation is via physicochemical and enzymological processes to separate the membranous layers of the stratum corneum from whole skin.[16,17] However, it is time consuming and, in some cases, difficult to control the size and thickness of a sheet of stratum corneum.

In these studies, the human callus (stratum corneum of the horny sole pads) was employed as *in vitro* model.[18] It is easier to obtain and prepare in powdered form (PHSC), and the particle sizes can be selected. Since a corneocyte is only about 0.5 μm thick and about 30 to 40 μm long, the PHSC is considered to contain intact corneocytes plus intercellular medium structures and to keep its physical–biochemical properties. Moreover, the exterior surface of the PHSC has been greatly extended and, therefore, favors solutes penetrating the PHSC from any direction.

This chapter has summarized a variety of potential applications for PHSC, ranging from basic science to applications in medicine and environmental concerns. PHSC, imagination, and a balanced study design can add to the scientific knowledge.

References

1. **Barry, B.W.,** Structure, function, diseases, and topical treatment of human skin, in *Dermatological Formulations: Percutaneous Absorption*, Barry, B.W., Ed., Marcel Dekker, New York, 1993, 1.
2. **Blank, J.H.,** Cutaneous barriers, *J. Invest. Dermatol.*, 45, 249, 1965.
3. **Scheuplein, R.J. and Blank, I.H.,** Mechanisms of percutaneous absorption. IV. Penetration of nonelectrolytes (alcohols) from aqueous solutions and from pure liquids, *J. Invest. Dermatol.*, 60, 286, 1973.
4. **Elias, P.M.,** Lipids and epidermal permeability barrier, *Arch. Dermatol. Res.*, 270, 95, 1981.
5. **Yardley, A.J.,** Epidermal lipids, *Int. J. Cosmet.*, 59, 13, 1987.
6. **Scheuplein, R.J. and Bronaugh, R.L.,** Percutaneous absorption, in *Biochemistry and Physiology of the Skin,* Vol. 1, Goldsmith, L.A., Ed., Oxford University Press, Oxford, 1993, 1255.
7. **Zatz, J.L.,** Scratching the surface: rational approaches to skin permeation, in *Skin Permeation Fundamentals and Application*, Zatz, J.L., Ed., Allured Publishing Co., Wheaton, IL, 1993, 11.
8. **Wester, R.C., Mobayen, M., and Maibach, H.I.,** *In vivo* and *in vitro* absorption and binding to powdered stratum corneum as methods to evaluate skin absorption of environmental chemical contaminants from ground and surface water, *J. Toxicol. Environ. Health*, 21, 367, 1987.
9. **Wester, R.C. and Maibach, H.I.,** Dermatopharmacokinetics in clinical dermatology, *Semin. Dermatol.*, 2, 81, 1983.
10. **Wester, R.C., Bucks, D.A.W., and Maibach, H.I.,** *In vivo* percutaneous absorption of hydrocortisone in psoriatic patients and normal volunteers, *J. Am. Acad. Dermatol.*, 8, 645, 1983.
11. **Wester, R.C., Maibach, H.I., Sedik, L., Melendres, J., Di Zio, S., and Wade, M.,** *In vitro* percutaneous absorption of cadmium from water and soil into human skin, *Fundam. Appl. Toxicol.*, 19, 1, 1992.
12. **Wester, R.C., Maibach, H.I., Sedik, L., Melendres, J., and Wade, M.,** *In vivo* and *in vitro* percutaneous absorption and skin decontamination of arsenic from water and soil, *Fundam. Appl. Toxicol.*, 20, 336, 1993.
13. **Guy, R.H. and Hadgraft, J.,** Principles of skin permeability relevant to chemical exposure, in *Dermal and Ocular Toxicology: Fundamentals and Methods*, Hobson, D.W., Ed., CRC Press, Boca Raton, FL, 1991, 221.
14. **Surber, C., Wilhelm, K.P., Hori, M., Maibach, H.I., and Guy, R.H.,** Optimization of topical therapy: partitioning of drugs into stratum corneum, *Pharmceut. Res.*, 7(12), 1320, 1990.
15. **Potts, P.O. and Guy, R.H.,** Predicting skin permeability, *Pharmaceut. Res.*, 9(5), 663, 1992.
16. **Juhlin, L. and Shelly, W.B.,** New staining techniques for the Langerhans cell, *Acta Dermatol. (Stockholm)*, 57, 289, 1977.

17. **Knutson, K., Potts, R.O., Guzek, D.B., Golden, G.M., Mckie, J.E., Lambert, W.J., and Higuchi, W.I.,** Macro and molecular physical-chemical considerations in understanding drug transport in the stratum corneum, *J. Contr. Rel.*, 2, 67, 1985.
18. **Hui, X., Wester, R.C., Magee, P.S., and Maibach, H.I.,** Partitioning of chemicals from water into powdered human stratum corneum (callus)—a model study, *In Vitro Toxicol.*, 8(2), 159, 1995.